Climate Change in Complex Systems

Climate Change in Complex Systems

Effects, Adaptations, and Policy Considerations for Agriculture and Ecosystems

Editors

Bruce A. McCarl
Anastasia W. Thayer
Thomas Lacher
Aurora M Vargas

MDPI • Basel • Beijing • Wuhan • Barcelona • Belgrade • Manchester • Tokyo • Cluj • Tianjin

Editors
Bruce A. McCarl
Texas A&M University
USA

Anastasia W. Thayer
Texas A&M University
USA

Thomas Lacher Texas
A&M University USA

Aurora M Vargas
Texas A&M University
USA

Editorial Office
MDPI
St. Alban-Anlage 66
4052 Basel, Switzerland

This is a reprint of articles from the Special Issue published online in the open access journal *Climate* (ISSN 2225-1154) (available at: https://www.mdpi.com/journal/climate/special_issues/climate_agriculture_ecosystems).

For citation purposes, cite each article independently as indicated on the article page online and as indicated below:

LastName, A.A.; LastName, B.B.; LastName, C.C. Article Title. *Journal Name* **Year**, *Article Number*, Page Range.

ISBN 978-3-03936-942-3 (Hbk)
ISBN 978-3-03936-943-0 (PDF)

Cover image courtesy of Thomas Lacher.

Contents

About the Editors

Bruce A. McCarl MCCARL is a University Distinguished Professor, Presidential Impact Fellow, Regents Professor, Senior AgriLife Research Fellow and Professor of Agricultural Economics at Texas A&M. Dr. McCarl joined Texas A&M in 1985, having previously worked at Oregon State and Purdue. He earned a B.S. in Business Statistics from the University of Colorado and a Ph.D. in Management Science from Pennsylvania State University. His areas of interest are the economic implications of the food–energy–water nexus; global climate change and greenhouse gas emission reduction; forestry and agricultural policy design; biofuels; mathematical programming, and risk analysis. He is the author of 296 journal articles and over 500 other papers and presentations. He has been involved sponsored research amounting to over \$82 million. He is a Fellow of the Agricultural and Applied Economics Association and a Fellow of both the Western and Southern Agricultural Economics Associations. He was a member of the Intergovernmental Panel on Climate Change that was co-recipient of the 2007 Nobel Peace Prize.

Anastasia W. Thayer is an agricultural economist with a focus on natural resources. She is currently an Assistant Professor in the Department of Applied Economics at Utah State University. Her previous projects cover the topics of climate change impacts in agriculture, water use in agriculture, and the ways in which water markets can change water allocation among user groups. She received her PhD in Agricultural Economics from Texas A&M University and her M.S. in Resource Economics from the University of Alaska Fairbanks.

Thomas Lacher is a Full Professor in Ecology and Conservation Biology at Texas A&M University and Director of the Center for Coffee Research and Education at the Borlaug Institute. He has held positions at the University of Brasilia, Brazil, Western Washington University, and Clemson University, where he was the executive director of the research consortium of the Archbold Tropical Research Center. From 2002 to 2007, he was Senior Vice-President and Executive Director of the Center for Applied Biodiversity Science at Conservation International, leading projects focused on conservation and sustainable development around the globe. At Texas A&M University, he was Professor and Caesar Kleberg Chair in Wildlife Ecology in the Department of Wildlife and Fisheries Sciences (1996–2002) and was also Head of the Department from 2007 to 2011. Dr. Lacher has been working in the tropics for over 40 years, with field research experience in Dominica, Mexico, Costa Rica, Panama, Colombia, Guyana, Suriname, Peru, and Brazil. From 2013 to 2017, he was co-PI on USAID/Uganda Environmental Management for the Oil Sector SOL-617-12-000026. He is an Associate Conservation Scientist at the Global Wildlife Conservation and is a member of the IUCN Climate Change Specialist Group, Co-Chair of the IUCN Small Mammal Specialist Group, and he serves on the IUCN Red List Committee.

Aurora M Vargas is an agricultural economist with an interest in data science and the analysis of diverse data sets. She received her B.S. in Animal Science from Louisiana State University and her Ph.D. in Agricultural Economics from Texas A&M University. Throughout her doctoral program, she researched food–water–energy nexus goals and established alternative strategies for improved management. Her publications can be found in the Climate and Energy Proceedings journals. Currently, Dr. Vargas is working as a quantitative analyst consultant focused on the correct implementation of data and modelling towards determining the individual outcomes of the United States horse racing industry

Editorial

Disconnect within Agriculture and Ecosystem Climate Effects, Adaptations and Policy

Anastasia W. Thayer [1,*], Aurora M Vargas [2], Thomas E. Lacher [3,4] and Bruce A. McCarl [5]

1   Department of Applied Economics, Utah State University, Logan, UT 84322, USA
2   Agricultural Economics/College of Agriculture, Texas A&M University, College Station, TX 77843, USA; avarga5@tamu.edu
3   Department of Wildlife and Fisheries Sciences, TexasA&M University, College Station, TX 77843, USA; tlacher@tamu.edu
4   Center for Coffee Research and Education, Texas A&M University, College Station, TX 77843, USA
5   Department of Agricultural Economics, Texas A&M University, College Station, TX 77843, USA; brucemccarl@gmail.com
*   Correspondence: anastasia.thayer@usu.edu

Received: 30 April 2020; Accepted: 8 May 2020; Published: 13 May 2020

1. Introduction

Frequently, agriculture and ecosystems (AE) are seen as separate entities, causing entity specific solutions in response to threats. Anthropogenic climate change simultaneously stresses both agriculture and ecosystems along with their interactions. Induced increasing surface temperatures [1], altered precipitation [2], drought intensification [3], altered ground and surface water quantity/quality [4,5], and diminished soil moisture [6] force adaptations for AE, but these adaptations fail to be efficient when interdependencies are not considered. Additional adaptations will be necessary, as future projections anticipate even greater climate change [1].

Research has quantified many AE impacts of climate change and yet greater impacts are anticipated as climate change proceeds. Thus, understanding the implications for AE systems is crucial. AE function, health, and productivity depend heavily on climatic characteristics. Typically, agriculture gets the most attention, as it feeds the world; however, an adaptation that only considers agriculture can negatively affect ecosystems and vice versa. Failure to incorporate the overlapping effects of agriculture and ecosystems could lead to maladaptation and greater long-term damages under climate change. The papers in this issue address a number of aspects of this issue.

Table 1 is adapted from Thayer et al., 2020 [7] and it provides examples of external ecological effects of agricultural focused adaptations and vice versa. Column 1 displays the general climate stressor with Column 2 showing the particular effect that has been seen in select areas. Columns 3 and 4 show either agricultural adaptations and their unintended impact on the ecosystem [termed an externality] or an ecosystem adaptation with the unintended result on agriculture [termed externality].

The examples demonstrate how an adaptation in agriculture or ecosystems can impact the other. Another factor to keep in mind is that climate change and its effects vary across the landscape geography as does AE characteristics; thus, adaptation actions must address local AE situations and cannot be spatial uniform.

This editorial will review the collective findings in the papers that are published in the *Climate* Special Issue "Climate Change in Complex Systems: Effects, Adaptations, and Policy Considerations for Agriculture and Ecosystems". We will discuss the ways the papers address climate change impacts, potential adaptations, and future policy for the continued AE prosperity. We also discuss the identified needs for research and future directions of AE interface adaptation research.

Table 1. Adaptations and externalities in response to climate stressors and effects, adapted from [7]

Climate Stressor	Climate Effect	Agricultural Adaptation	Ecosystem Service Externality
Increased temperature and drought	**Increased livestock heat stress and reduced forage and growth** [8]	Diversifying livestock species [9–11]	Altered plant biodiversity and productivity [12–14]
	Lower crop production and quality due to increased temperatures affecting growth and nutrient content [15,16]	Crop land shift to grazing [17–19]	Increased root production in upper soil levels and carbon sequestration [20,21].
Climate Stressor	**Climate Effect**	**Ecosystem Adaptation**	**Agricultural System Externality**
Increased drought	**Reduced plant growth** due to changes in temperature, precipitation, or the incidence of climatic extremes [22,23]	Shift in vegetation mix productivity and water retention [24,25]	Altered water supply and increased demand for irrigation [26,27]
Increased temperature and altered rainfall	**Disruption in Hydrological environments** that cycle nutrients, maintain water quality, and moderatelifecycle events such as spawning and recruitment [28–31]	Shifting species distribution including pest incidence [32,33]	Increased pesticide and herbicide costs [34–36]

2. Comments on Effects

Regions experience differential impacts and researchers have used diverse methods to quantify climate change effects on AE due to the complex nature of climate. Every paper in the special issue clearly identifies current and future climate change impacts on their study area.

Sinay and Carter (2020) reviewed papers that focused on climate effects on coastal communities [37]. They discussed climate change as a cause of increased occurrences of flooding and fire along with the impacts to coastlines and beaches, inland areas, infrastructure, housing, natural systems, food production, fresh and drinking water availability, and community welfare.

Changes in water availability and use is expected under climate change and has been observed to have varying impacts on AE systems within the special issue. Elijah and Odiyo show that Kenyan droughts have increased the use of groundwater to sustain rainfed agriculture, which leads to increased soil salinity due to irrigation [38]. Scholes illustrates that South Africa is also experiencing land degradation, due to high solar radiation, low atmospheric humidity and rainfall, and increased seasonality and variability of rainfall, causing a shift away from animal production and potentially to energy production [39]. Scholes (2020) further highlighted that semi-arid regions will be particularly vulnerable to land degradation and an expansion of desertification. In the paper by Ngarava et al., South Africa is also struggling to increase its livestock and energy production under climatic stressors while attempting to reduce carbon dioxide emissions [40].

Further, water stress and increased temperatures were discussed in various regions in Korea and the United States. An et al. report increased insect populations as a result of rising temperatures and decreased tree health due to water stress are affecting the growth of the Korean Oak and, in turn, the country's lumber industry [41]. In addition, Ding and McCarl show that, under increased drought, a region of Texas with competing interests in water rights is expected to experience crop losses and a shift from expensive irrigated land to grasslands [42]. Further, as groundwater pumping for municipal and industrial water increases, lower pumping limits might be imposed, which could jeopardize the ecosystems that rely on the spring levels fed by the groundwater systems.

As discussed, climate effects may have common aspects across the landscape, but their solutions will require localized attention and they are subject to available resources, magnitude and knowledge of current and future impacts, as well as the community's response. Thus, a collection of viable adaptations must be outlined to facilitate and lessen the expected damage as a result of climate effects.

3. Comments on Adaptation

Identifying appropriate adaptations was a key goal in designing this special issue. However, few papers in this collection suggested specific AE adaptation strategies. Only Sinay and Carter exclusively focused on identifying and synthesizing the best practices in adaptation strategies [37]. Other papers were able to make adaptation suggestions specific to the system such as Scholes argument for the adoption of sustainable land use [39] or Ding and McCarl's suggested changes to current water use [42].

However, none of the studies were able to fully discuss adaptations in the context of both ecosystems and agriculture.

Despite a lack of concrete adaptations for each system, other take-aways from the literature might be relevant when suggesting future productive directions for adaptation research. In general, Sinay and Carter suggest that adaptation strategies should be flexible and multiple strategies might need to be considered in order to respond to the magnitude of effects [37]. Identifying a range of possible adaptations or a time frame where one adaptation might be more effective could be productive. Several of the papers cited here were also able to identify adaptations that might not be useful [37–39,41]. While the scope of study areas and methodologies suggests that adaptations discussed in these papers are difficult to summarize, it might be helpful for future research to discuss adaptations that are likely to lead to maladaptation or worse outcomes just as much as suggest adaptations.

It is known that identifying adaptation strategies is difficult and their role to combat the effects of future changes is complex [43]. Despite this difficulty, climate change impact studies have insights into the study region, knowledge of the drivers, which impact the magnitude of effects, and an understanding of system feedbacks. These factors will be critical in estimating the magnitude of future effects and identifying best adaptation practices that benefit, or do not worsen, the agriculture and ecosystems. Thus, future research studies must extend their scope to consider adaptation strategies for the effects that they present as key findings. This could include drawing on literature from other similar study areas, as did Scholes [39], or attempting to extend the analysis and discussion to explicitly extend the findings from one system (agriculture or ecosystems) to discussing adaptations that will be necessary in other systems [7].

4. Comments on Policy

While papers that were included in this special edition fell short of providing concrete adaptation strategies that addressed AE simultaneously, studies were more successful in identifying policy recommendations to respond to current and future climate change effects; however, papers fell short of calling these policies adaptation strategies.

Policy recommendations were generally specific to the particular study area and they emphasized the need for local solutions and investments in human capital, such as the recommendation of several papers on education for success [37–39]. It was also clear that, if properly designed, financial incentives and economic support mechanisms could be useful in a number of study areas [40,41]. Ding and McCarl were able to point to specific policy recommendations and their impact on the community and discuss the effects of a policy on both humans and the ecosystem [42].

The contrast between authors' ability to make policy recommendations and suggest adaptation strategies suggests a possible important disconnect in researchers' ability and confidence in discussing the future impacts of climate change. In general, the distinction between policy recommendations and adaptations seemed to be arbitrary and only delineated by the timeframe the policy would be put in place. In many cases, policy recommendations were framed as such and not as adaptations to climate change. This might highlight the need for education of climate change researchers to adaptation scenarios and their ability to restructure research topics in order to explore adaptations. In many cases, with slight augmentation of research or extensions, policy recommendations could be easily tested as either successful or unsuccessful adaptations to climate change effects. Extending research to include a formal explanation and discussion of adaptation strategies reduces the risk to the study area and provides tested best-responses.

5. Conclusions

This special edition attracted a diverse selection of papers that were focused on climate change effects, adaptations, and policy recommendations with the goal of exploring agriculture and ecosystems impacts and interdependencies. As noted, the broad range in scope made it difficult to make concrete conclusions across each area of focus: effects, adaptations, and policy. Further, while the authors attempted to blend ecosystems and agriculture into a holistic sphere of research, largely, this remains a difficult and incomplete objective. This suggests that the field of climate change research in the AE arena needs additional support, funding, and ways to prioritize and incentivize integrated research and interdisciplinary teams in order to generate findings that will be applicable and accurate to the complex systems that define reality [7].

From the wide scope of articles included in this collection, it is clear that how humans and ecosystems respond to climate change effects will have a large influence on the eventual impact of changes. In all papers, land use changes in the coming decades, resource use, and conservation efforts, as well as energy use and efficiency efforts will define the ultimate failure or success of governmental and institutional responses to climate change as we transgress into the Anthropocene [44].

Conflicts of Interest: The authors declare no conflict of interest. The funders had no role in the design of the study; in the collection, analyses, or interpretation of data; in the writing of the manuscript, or in the decision to publish the results.

References

1. IPCC. *Climate Change 2014: Impacts, Adaptation, and Vulnerability. Part A: Global and Sectoral Aspects. Contribution of Working Group II to the Fifth Assessment Report of the Intergovernmental Panel on Climate Change*; Field, C.B., Barros, V.R., Dokken, D.J., Mach, K.J., Mastrandrea, M.D., Bilir, T.E., Chatterjee, M., Ebi, K.L., Estrada, Y.O., Genova, R.C., et al., Eds.; Cambridge University Press: Cambridge, UK; New York, NY, USA, 2014.
2. Van Vliet, M.T.H.; Franssen, W.H.P.; Yearsley, J.R.; Ludwig, F.; Haddeland, I.; Lettenmaier, D.P.; Kabat, P. Global river discharge and water temperature under climate change. *Glob. Environ. Chang. Hum. Policy Dimens.* **2013**, *23*, 450–464. [CrossRef]
3. Trenberth, K.; Dai, A.; van der Schrier, G.; Jones, P.H.; Barichivich, J.; Briffa, K.R.; Sheffield, J. Global warming and changes in drought. *Nat. Clim. Chang.* **2014**, *4*, 17–22. [CrossRef]
4. Taylor, R.; Scanlon, B.; Döll, P.; Rodell, M.; van Beek, R.; Wada, Y.; Longuevergne, L.; Leblanc, M.; Famiglietti, J.S.; Edmunds, M.; et al. Groundwater and climate change. *Nat. Clim. Chang.* **2013**, *3*, 322–329. [CrossRef]
5. Whitehead, P.G.; Wilby, R.L.; Battarbee, R.W.; Kernan, M.; Wade, A.J. A review of the potential impacts of climate change on surface water quality. *Hydrol. Sci. J.* **2009**, *54*, 101–123. [CrossRef]
6. Derner, J.D.; Johnson, H.B.; Kimball, B.A.; Pinter, P.J.; Polley, H.W.; Tischler, C.R.; Boutton, T.W.; LaMorte, R.L.; Wall, G.W.; Adam, N.R.; et al. Above-and below-ground responses of C3–C4 species mixtures to elevated CO2 and soil water availability. *Glob. Chang. Biol.* **2003**, *9*, 452–460. [CrossRef]
7. Thayer, A.W.; Vargas, A.; Castellanos, A.A.; Lafon, C.W.; McCarl, B.A.; Roelke, D.L.; Winemiller, K.O.; Lacher, T.E. Integrating Agriculture and Ecosystems to Find Suitable Adaptations to Climate Change. *Climate* **2020**, *8*, 10. [CrossRef]
8. Rötter, R.; Van de Geijn, S.C. Climate change effects on plant growth, crop yield and livestock. *Clim. Chang.* **1999**, *43*, 651–681. [CrossRef]
9. Seo, S.N.; McCarl, B.A.; Mendelsohn, R.O. From beef cattle to sheep under global warming? An analysis of adaptation by livestock species choice in South America. *Ecol. Econ.* **2010**, *69*, 2486–2494. [CrossRef]
10. Zhang, Y.W.; McCarl, B.A.; Jones, J.P.H. An Overview of Mitigation and Adaptation Needs and Strategies for the Livestock Sector. *Climate* **2017**, *5*, 95. [CrossRef]
11. Seo, S.N.; Mendelsohn, R.O.; Dinar, A.; Kurukulasuriya, P. Adapting to climate change mosaically: An analysis of African livestock management by agro-ecological zones. *B.E. J. Econ. Anal. Policy* **2009**, *9*. [CrossRef]
12. Fuhlendorf, S.D.; Engle, D.M. Restoring heterogeneity on rangelands: Ecosystem management based on evolutionary grazing patterns: We propose a paradigm that enhances heterogeneity instead of homogeneity to promote biological diversity and wildlife habitat on rangelands grazed by livestock. *BioScience* **2001**, *51*, 625–632. [CrossRef]
13. Megersa, B.; Markemann, A.; Angassa, A.; Ogutu, J.O.; Piepho, H.P.; Zárate, A.V. Livestock diversification: An adaptive strategy to climate and rangeland ecosystem changes in southern Ethiopia. *Hum. Ecol.* **2014**, *42*, 509–520. [CrossRef]

14. Pequeño-Ledezma, M.; Alanís-Rodríguez, E.; Molina-Guerra, V.M.; Mora-Olivo, A.; Alcalá-Rojas, A.G.; Martínez-Ávalos, J.G.; Garza-Ocañas, F. Plant composition and structure of two post-livestock areas of Tamaulipan thornscrub, Mexico. *Rev. Chil. Hist. Nat.* **2018**, *91*, 1–8. [CrossRef]
15. Derner, J.; Briske, D.; Reeves, M.; Brown-Brandl, T.; Meehan, M.; Blumenthal, D.; Travis, W.; Augustine, D.; Wilmer, H.; Scasta, D.; et al. Vulnerability of grazing and confined livestock in the Northern Great Plains to projected mid-and late-twenty-first century climate. *Clim. Chang.* **2018**, *146*, 19–32. [CrossRef]
16. Craine, J.M.; Elmore, A.; Angerer, J.P. Long-term declines in dietary nutritional quality for North American cattle. *Environ. Res. Lett.* **2017**, *12*, 044019. [CrossRef]
17. Mu, J.E.; McCarl, B.A.; Wein, A.M. Adaptation to climate change: Changes in farmland use and stocking rate in the U.S. *Mitig. Adapt. Strateg. Glob. Chang.* **2013**, *18*, 713–730. [CrossRef]
18. Cho, S.J.; McCarl, B.A. Climate change influences on crop mix shifts in the United States. *Sci. Rep.* **2017**, *7*, 40845. [CrossRef]
19. Joyce, L.A.; Briske, D.D.; Brown, J.R.; Polley, H.W.; McCarl, B.A.; Bailey, D.W. Climate change and North American rangelands: Assessment of mitigation and adaptation strategies. *Rangel. Ecol. Manag.* **2013**, *66*, 512–528. [CrossRef]
20. Derner, J.D.; Boutton, T.W.; Briske, D.D. Grazing and ecosystem carbon storage in the North American Great Plains. *Plant Soil.* **2006**, *280*, 77–90. [CrossRef]
21. Derner, J.D.; Schuman, G.E. Carbon sequestration and rangelands: A synthesis of land management and precipitation effects. *J. Soil Water Conser.* **2007**, *62*, 77–85.
22. Allen, C.D.; Macalady, A.K.; Chenchouni, H.; Bachelet, D.; McDowell, N.; Vennetier, M.; Kitzberger, T.; Rigling, A.; Breshears, D.D.; Hogg, E.H.; et al. A global overview of drought and heat-induced tree mortality reveals emerging climate change risks for forests. *For. Ecol. Manag.* **2010**, *259*, 660–684. [CrossRef]
23. Aragón-Gastélum, J.L.; Flores, J.; Yáñez-Espinosa, L.; Badano, E.; Ramírez-Tobías, H.M.; Rodas-Ortíz, J.P.; González-Salvatierra, C. Induced climate change impairs photosynthetic performance in *Echinocactus platyacanthus*, an especially protected Mexican cactus species. *Flora* **2014**, *209*, 499–503. [CrossRef]
24. Geruo, A.; Velicogna, I.; Kimball, J.S.; Du, J.; Kim, Y.; Colliander, A.; Njoku, E. Satellite-observed changes in vegetation sensitivities to surface soil moisture and total water storage variations since the 2011 Texas drought. *Environ. Res. Lett.* **2017**, *12*, 054006. [CrossRef]
25. Schwantes, A.M.; Swenson, J.J.; González-Roglich, M.; Johnson, D.M.; Domec, J.C.; Jackson, R.B. Measuring canopy loss and climatic thresholds from an extreme drought along a fivefold precipitation gradient across Texas. *Glob. Chang. Biol.* **2017**, *23*, 5120–5135. [CrossRef] [PubMed]
26. McDonald, R.I.; Girvetz, E.H. Two Challenges for U.S. Irrigation Due to Climate Change: Increasing Irrigated Area in Wet States and Increasing Irrigation Rates in Dry States. *PLoS ONE* **2001**, *8*, e65589. [CrossRef] [PubMed]
27. Rodríguez-Díaz, J.A.; Weatherhead, E.K.; Knox, J.W.; Camacho, E. Climate change impacts on irrigation water requirements in the Guadalquivir river basin in Spain. *Reg. Environ. Chang.* **2007**, *7*, 149. [CrossRef]
28. Richter, B.D. Ecologically sustainable water management: Managing river flows for ecological integrity. *Ecol. Appl.* **2003**, *13*, 206–224. [CrossRef]
29. Perkin, J.S.; Gido, K.B.; Costigan, K.H.; Daniels, M.D.; Johnson, E.R. Fragmentation and drying ratchet down Great Plains stream fish diversity. *Aquat. Conserv. Mar. Freshw. Ecosyst.* **2015**, *25*, 639–655. [CrossRef]
30. Postel, S.; Carpenter, S. Freshwater ecosystem services. In *Nature's Services: Societal Dependence on Natural Ecosystems*; Daily, G.C., Ed.; Island Press: Washington, DC, USA, 1997; pp. 195–214. ISBN 9781559634762.
31. Durham, B.W.; Wilde, G.R. Influence of stream discharge on reproductive success of a prairie stream fish assemblage. *Trans. Am. Fish Soc.* **2006**, *135*, 1644–1653. [CrossRef]
32. Mainali, K.P.; Warren, D.L.; Dhileepan, K.; McConnachie, A.; Strathie, L.; Hassan, G.; Karki, D.; Shrestha, B.B.; Parmesan, C. Projecting future expansion of invasive species: Comparing and improving methodologies for species distribution modeling. *Glob. Chang. Biol.* **2015**, *21*, 4464–4480. [CrossRef]
33. Burlakova, L.E.; Karatayev, A.Y.; Karatayev, V.A.; May, M.E.; Bennett, D.L.; Cook, M.J. Biogeography and conservation of freshwater mussels (Bivalvia: Unionidae) in Texas: Patterns of diversity and threats. *Divers. Distrib.* **2011**, *17*, 393–407. [CrossRef]
34. Wolfe, D.W.; Ziska, L.; Petzoldt, C.; Seaman, A.; Chase, L.; Hayhoe, K. Projected change in climate thresholds in the Northeastern U.S.: Implications for crops, pests, livestock, and farmers. *Mitig. Adapt. Strateg. Glob. Chang.* **2008**, *13*, 555–575. [CrossRef]

35. Smith, R.G.; Menalled, F.D. *Integrated Strategies for Managing Agricultural weeds: Making Cropping Systems Less Susceptible to Weed Colonization and Establishment Department of Land Resources and Environmental Sciences*; Montana State University: Bozeman, MT, USA, 2006.

36. Chen, C.C.; McCarl, B.A. An investigation of the relationship between pesticide usage and climate change. *Clim. Chang.* **2001**, *50*, 475–487. [CrossRef]

37. Sinay, L.; Carter, R.W.B. Climate Change Adaptation Options for Coastal Communities and Local Governments. *Climate* **2020**, *8*, 7. [CrossRef]

38. Elijah, V.T.; Odiyo, J.O. Perception of Environmental Spillovers across Scale in Climate Change Adaptation Planning: The Case of Small-Scale Farmers' Irrigation Strategies, Kenya. *Climate* **2020**, *8*, 3. [CrossRef]

39. Scholes, R.J. The Future of Semi-Arid Regions: A Weak Fabric Unravels. *Climate* **2020**, *8*, 43. [CrossRef]

40. Ngarava, S.; Zhou, L.; Ayuk, J.; Tatsvarei, S. Achieving Food Security in a Climate Change Environment: Considerations for Environmental Kuznets Curve Use in the South African Agricultural Sector. *Climate* **2019**, *7*, 108. [CrossRef]

41. An, H.; Lee, S.; Cho, S.J. Climate Change Impacts on Forest Management: A Case of Korean Oak Wilt. *Climate* **2019**, *7*, 141. [CrossRef]

42. Ding, J.; McCarl, B.A. Economic and Ecological Impacts of Increased Drought Frequency in the Edwards Aquifer. *Climate* **2020**, *8*, 2. [CrossRef]

43. Tompkins, E.L.; Adger, W.L. Defining response capacity to enhance climate change policy. *Environ. Sci. Policy* **2005**, *8*, 562–571. [CrossRef]

44. Lacher, T.E., Jr.; Roach, N.S. The status of biodiversity in the Anthropocene: Trends, threats, and actions. In *Volume 3 (Biodiversity), the Encyclopedia of the Anthropocene*; Lacher, T.E., Jr., Pyare, S., Eds.; Elsevier: Oxford, UK, 2018; pp. 1–8. [CrossRef]

Article

Climate Change Adaptation Options for Coastal Communities and Local Governments

Laura Sinay * and R. W. (Bill) Carter

Sustainability Research Centre, School of Social Science, University of the Sunshine Coast, Sippy Downs Campus, 4558 Sunshine Coast, Australia; bcarter@usc.edu.au
* Correspondence: lsinay@usc.edu.au or laura.sinay@unirio.br; Tel.: +61-478048633

Received: 25 October 2019; Accepted: 18 December 2019; Published: 7 January 2020

Abstract: Extreme weather events and failure to adapt to the likely impacts of climate change are two of the most significant threats to humanity. Therefore, many local communities are preparing adaptation plans. Even so, much of what was done has not been published in the peer-reviewed literature. This means that consideration of adaptation options for local communities is limited. With the objective of assisting in the development of adaptation plans, we present 80 adaptation options suitable for coastal communities that can be applied by local governments. They are a catena of options from defend to co-exist and finally, retreat that progresses as impacts become less manageable. Options are organized according to their capacity to protect local properties and infrastructure, natural systems, food production, availability of fresh and drinking water and well-being of the local population, as these are likely to be affected by climate change. To respond to multiple threats, 'soft' options, such as awareness raising, planning, political articulation and financial incentives, insurance and professional skills enhancement, can be encouraged immediately at relatively low cost and are reversible. For specific threats, options emphasize change in management practices as pre-emptive measures. Key audiences for this work are communities and local governments starting to consider priority actions to respond to climate change impacts.

Keywords: climate change; adaptation; coastal community; local government; responses

---

1. Introduction

The Intergovernmental Panel on Climate Change (IPCC) reports that by 2100 anthropogenically-induced climate change is likely to lead to a rise in global temperatures of 1.5 °C to 2 °C above pre-industrial levels [1]. This will significantly affect environmental feedback systems leading to, among others, more frequent and intense extreme weather and climate-related events [1]. In this context, in 2019, the World Economic Forum ranked extreme weather events and failure to adapt to the likely impacts of climate change as the two most significant threats to humanity [2].

At the 2015 United Nations Framework Convention on Climate Change, Conference of the Parties (COP21) in Paris, 195 countries agreed to increase efforts to mitigate and adapt to climate change [3]. Mitigation of climate change refers to controlling the emission of greenhouse gases to retard the global warming process [4]. This is based on the understanding that temperature rise is directly related to the amount and type of greenhouse gases emitted into the atmosphere. Mitigation, therefore, refers to avoiding anthropogenically-induced climate change [4]. Despite the Paris agreement, the emission of greenhouse gases continues to increase [5] and, considering the political discourses of key countries, such as the US and Brazil, it is likely this pattern will continue in the years to come. As anthropogenically-induced climate change appears to be unavoidable [1], adaptation (the process of adjustment by which risks are managed to improve community safety and well-being [4]) becomes essential.

Risk stems from a combination of one or more threats and the capacity to respond to them [6]. A threat is something 'likely to cause damage or danger' [7]. For climate change, threats depend on how environmental feedback systems are affected. IPCC (2018) forecasts that by 2100, if global temperature rises (only) between 1.5 °C and 2 °C above pre-industrial levels, environmental feedback systems will lead to: extreme temperatures in many densely populated areas; more frequent and intense extreme weather and climate-related events, including droughts and floods; sea-level rise between 0.26–0.93 m; increased ocean acidity and de-oxygenated oceanic waters; and significant biodiversity loss [1]. The consequences of these global scale changes will be profound. The availability of potable water is likely to be affected by extended drought periods and the intrusion of seawater into inland waterways (caused by sea level rise and storm surge in coastal areas) [8]. Terrestrial and freshwater ecosystems will be affected by drought, flood, intrusion of seawater and change in temperatures [1,8]. Marine ecosystems will suffer through increased ocean acidity and water temperature and decreased oxygen levels, which are predicted to cause the loss of 70 per cent to 99 per cent of coral reefs [8]. A significant decline in biodiversity is predicted and likely to include local loss of pollinators, which with other threats, will put at risk food production [1]. Community well-being is expected to be affected by higher temperatures, more frequent and extreme weather events, and sea-level rise and storm surge will more frequently cause flooding in coastal and low-lying areas resulting in damage to infrastructure and properties [1]. How society responds to the forecasted risks is, therefore, paramount to the success of short and long-term sustainable development, community resilience [9] and resultant community well-being.

Despite the sensibility of responding to the threats of climate change through strategic and planned adaptive actions, much of what has been done lacks critical assessment in the peer-reviewed literature [10]. This means that appraisal of adaptation options for local communities is limited, and communities may take actions that are not best practice, and may be expensive, lack efficacy and be maladaptive [11]. Identification of adaptation options for local communities, councils and/or local industries is the first step in strategically responding to the threats of climate change to reduce risk to issues of concern, and the motivation for this study. Focus is on coastal communities, because they are particularly vulnerable to climate change impacts [11]. In addition, about 10 per cent of the World's population "live on coastal areas that are less than 10 m above sea level" [12]. Eighty adaptation options were identified as suitable for coastal communities and can be applied by local governments. With the objective of assisting with the development of adaptation plans, these options are described and discussed in the context of the broad adaptation options of retreat, co-exist and defend.

2. Materials and Methods

Adaptation options were first identified via a systematic literature review. Systematic reviews identify articles using clearly defined search criteria, and systematic, explicit and reproducible methods to select and critically examine relevant literature [13,14]. This approach is common in the health sciences and has been applied increasingly to environmental and climate change studies [15,16].

The peer-reviewed literature was systematically searched using Scopus©. Keywords used in the search were: climate change, adaptation, coastal, sea level rise, local government and storm surge ("TITLE-ABS-KEY ("climate change" AND adaptation AND coastal AND "sea level rise" OR "storm surge") AND DOCTYPE (ar) AND PUBYEAR > 2009"). Articles not in English, published prior to 2010, and book reviews were excluded. The Scopus search retrieved 114 results. These were analyzed and works that did not directly mention adaptation options were excluded. Based on this criterion, 44 works were selected for further analysis and reviewed in full.

Adaptation options identified were categorized in tables according to focus and response to threats (multiple threats, property and infrastructure, coastal flooding, inland flooding, fire, natural systems, farming, fresh and drinking water and well-being of communities). It was then noted that some category components (e.g., housing) lacked implicit adaptation options. In these cases, additional information was sourced from technical reports.

Because local demographic characteristics and cultural systems play important roles in defining adaptation options, a complete list of responses specific to a locale is unlikely to be identifiable from the literature. Hence, what we present here are widely applicable adaptation options found in the peer-reviewed literature plus those found in other sources that are likely to guide adaptation.

3. Results

3.1. Broad Adaptation Options: Retreat, Co-Exist and Defend

Climate change adaptation, disaster risk reduction and hazard assessment are strongly linked [17], and it is evident that this intersection is a strategic planning challenge for coastal communities [18]. Herein, we focus on the role of climate change on flood hazards, sea level rise, storm surges/cyclones and coastal erosion, as well as the interactions of these climate change affected hazards. It is in this nexus of climate change adaptation and disaster risk reduction where some of the major challenges exist for local governments and planners (e.g., [18]).

The focus on 'defend, co-exist, or retreat' is closely aligned with the define-analyze-implement-reassess (DAIR) framework developed as a general community and rural planning template for hazards affected by climate change [18]. Our study evaluated three possible strategies for decision-making related to coastal development and management of existing coastal resources: (1) relocate assets and people to safe areas (retreat); (2) defend existing and new structures against climate change affected hazards using largely structural measures; and (3) co-exist or adapt to changing conditions by a combination of innovative planning measures [18,19].

3.1.1. Retreat

Retreat, as the name suggests, refers to moving communities, structures and/or assets from areas that are likely to be significantly affected by the impacts of global warming to areas less vulnerable to climate change impacts [20]. This can be the case, for example, of inhabited areas expected to experience increased frequency and extent of flooding and storm surge. While it is a radical approach, it seems that more than one million people have already been forced to retreat in response to natural hazards [21,22]. While retreat is taking place in at least 27 different places, it is not an easy option [23]. In addition to the obvious cost of the land, which in many cases is bought by the government, and of reallocation, it has psychological and socio-cultural implications that, in many cases, cannot be mitigated [21,22].

Examples of retreat include Native Americans on the Isle de Jean Charles in Louisiana off the Gulf of Mexico, who collectively decided to retreat in 2016; inhabitants of Grantham in Queensland, Australia; and in Oakwood Beach (New Jersey) on the fringe of New York City [12].

The community of Belongil Beach, in Byron Bay (Australia) adopted a managed retreat approach. The decision was based on the understanding that "infrastructure, private property and residential development are located within the coastal erosion 'immediate hazard zone', which is the area of shoreline predicted to erode as the result of a 100-year average recurrence interval (ARI) design storm". The erosion areas were identified and classified in accordance to the expected time for erosion. Based on this, different zonings were established: "Immediate coastal hazard precinct: buildings are to be entirely modular and relocatable (by 4WD vehicle); trigger distance for relocation of development is 20 m from the coastal erosion escarpment; no building is to be within 20 m of the erosion escarpment. 50-year precinct: all residential housing is to be relocatable (by 4WD vehicle); trigger distance for relocation and/or demolition of development is 50 m from the erosion escarpment. 100-year precinct: trigger distance for relocation and/or demolition of development is 50 m from the erosion escarpment" [23].

3.1.2. Defend

Defend includes strategies implemented to protect assets from the impacts of flooding. It can involve the construction of seawalls and reforestation of riparian areas [20]. Defend is the most common response, as its disadvantages tend to be limited to financial costs, which can significantly vary depending on the circumstances. However, due to multiple reasons (e.g., storm surge maybe more intense than forecasted), defense tends to eventually fail; in which case either retreat or co-existence will be the options of choice.

Saibai, which is one of the nearly 300 islands that constitute the Torres Strait Islands archipelago (Australia), is an example of a locality where the local population had to implement defensive measures [24]. In 2016, it had an estimated population of 465 people, 85.6% of whom were Torres Strait Islander or Aboriginal. The 108 km^2 island has an average elevation of one meter with its highest point being 1.7 m above mean sea level [25]. Therefore, it is prone to flooding, especially during the wet season, which coincides with the cyclone season and king tides [26]. The combination of increasing mean sea level rise twice that of the global rate, coastal erosion and extreme weather events leave only defensive responses other than the option to retreat. Due to the high cost associated with building a seawall around the island, the first response was to use sand bags [27]. As the situation worsened, the community built a handcrafted seawall that was substituted in 2017 by a Government built seawall [28]. Half a year after the $24.5 million Government built seawall was inaugurated, it was breached by a high tide [29].

3.1.3. Co-Exist

Co-existence with climate change related threats refers to coping with the new conditions [30]. This option is based on the understanding that, while in specific situations nothing can be done to mitigate certain impacts, retreat and defend are either not necessary, excessively costly, or not possible. The option involves acceptance of losses, and communities must accept the risks of climate change and respond intermittently to the effects. Co-existence, besides being an expensive option in monetary terms, can be extremely traumatic due to the exposure to crisis events that may include loss of life. Despite its disadvantages, at present it appears to be the option most commonly adopted.

Examples in Australia of co-existence with threats imposed by climate change include towns such as Townsville and Cairns, which are intermittently but significantly affected by cyclone and floods, and Brisbane, which is periodically affected by flooding of the Brisbane River.

Cyclone Yasi, for example, caused extensive damage from Cooktown to Townsville. It almost destroyed everything in its path, including 9000 km of roads and 4500 km of rail, crops, houses and businesses [31]. Restoration costs, from just this one cyclone, was around $7 billion [31].

Brisbane floods are also frequent. Substantial floods, during which one-third or more of Brisbane city was inundated, occurred in 1841, 1844, 1890, 1893, 1898, 1974 and January 2011. Reconstruction of damage caused by floods costs some $100 million per year [32]. In addition to financial costs, the 1893 flood was associated with the loss of 35 lives, the 1974 flood with 14 deaths and more than 300 people injured; and the 2011 flood included 33 deaths (three others are still missing). The reconstruction of Brisbane from just this last flood exceeded $5 billion [18,32].

3.2. Punctuated Adaptation Options

The broad adaptation options of retreat, co-exist and defend are possible for all identified threats associated with climate change with efficacy varying with specific threats and circumstances [33,34]. Within these, the literature identifies specific adaptive actions that reflect defensive or co-existing strategies that fall short of the ultimate strategy of retreat. The actions tend to be presented as discrete or punctuated choices without consideration of comparative efficacy, synergies, or priority.

3.2.1. Adaptation Options for Responding to Multiple Threats

While some adaptation options are limited to address just one challenge, others are likely to have a systemic ameliorating effect and can be subdivided in five main approaches (Table 1).

Table 1. Options for responding to multiple threats.

Adaptation Options	References
Education and public awareness (co-exist)	[6,35–38]
Community participation (co-exist)	[20,36,38–40]
Integration between different government levels and financial incentives (co-exist)	[17,40,41]
Labor and professional skills enhancement (co-exist)	[10,42]
Flooding intensity map (co-exist)	[41]
Planning and legislation (co-exist)	[20,36,40–46]
Warning systems (co-exist)	[6,20,35]
Disaster management plan and evacuation plans (co-exist)	[35,41]

- Raise public awareness to encourage the local population to adapt and be prepared for the likely impacts of climate change and to foster community participation in decision-making.
- Planning (1) to avoid the worst consequences of the forecasted weather events, which involves production and frequent update of flooding and hazard maps; (2) legislation focused on (a) not permitting development on land vulnerable to hazards, or (b) establishing construction codes appropriate for the forecasted climatic conditions; and (3) plans to rapidly and efficiently respond to disasters, with disaster management and evacuation plans and warning systems to alert residents about imminent threats, such as fire and flooding.
- Political articulation and financial incentives to facilitate the integration between different government levels (regional, state and national) so that complex and expensive responses (e.g., rapid evacuation and/or construction of dykes) can be put in place in a timely manner. Financial incentives need to be also focused on both residents and industries so that economic constraints are not the main obstacle to allowing adaptation options to be implemented.
- Insurance to cover personal and government costs specifically associated with recovery after, for example, inundation, fire, weather events and/or failed crops.
- Labor and professional skills adjustment and enhancement so that new construction codes and new farming standards can be put in place.

3.2.2. Adaptation Options for Protecting Property and Infrastructure

In the context of risks associated with climate change, the main threats to properties and infrastructure on coastal areas are: flooding caused by sea level rise, storm surge, tide, freshwater flooding and wave run-up [47]; as well as fire, especially during droughts.

(1) Flooding

Flooding is already a problem in many places and it is likely to be aggravated with future sea level rise and with the intensification of storms. While inundation will mostly affect the coastline, it can also affect inland low-lying areas. If no adaptation options are put on place, property losses can be expected.

• Coastlines and beaches

For the protection of the coastline, identified adaptation options included two main approaches (Table 2). The first is construction of physical barriers (e.g., seawalls, breakwaters, gabion, groins and sluices). This response type tends to squeeze the intertidal habitat, resulting in a reduction in habitat and a usurpation of the natural resilience of the habitat usually because of poor understanding of structural and ecological dynamics. Consequentially, but with an anthropocentric bias, this response should only be initiated "where erosion presents an imminent threat to public safety or infrastructure that cannot practicably be removed or relocated. Where erosion protection structures are necessary, maintaining physical coastal processes outside the area subject to the coastal protection works is required to avoid adverse impacts on adjacent coastal landforms and associated ecosystems" [48].

Table 2. Options for responding to coastal fringe flooding.

Adaptation Options	References
Construction of physical barriers	
Seawalls, breakwaters, gabion, groins and sluices (defend)	[30,35,41,44,49–54]
Environmental management	
Protection of mangrove, wetlands, dunes forests and reforestation of areas close to waterways (defend)	[30,38,41,44,50,52,55,56]
Creation of artificial reefs (defend)	[57]
Prohibition or control of the removal of beach sediments (defend)	[20]
Beach nourishment (defend)	[44,50,58]

The second is less construction oriented and involves improved environmental management, with approaches such as (a) protection of existing ecosystems and reforestation of areas adjacent to coastlines to reduce flooding from storm surge and dissipate the energy of waves, and hence, lessen the impact of wave run-ups; (b) beach nourishment focused on maintaining coastlines at a predetermined width (This can "disrupt species living, feeding, and nesting on the beach", as well as the habitats at dredging sites; also "it is infeasible in areas where the wave energy is very high" [50,55]. Despite the likely negative impacts, beach nourishment is a common and frequent practice in many coastal areas.); (c) prohibition or control of the removal of beach sediments, because removal may accelerate beach erosion and disturb fauna; and (d) creation of artificial reefs to dissipate wave energy (and help to support marine biota).

- Inland areas

Adaptation options for avoiding inland inundation include elevating existing or constructing new canals and river walls, dykes, or sluices (Table 3). These options can be costly and if not planned for the highest predicted flood levels may prove ineffective. Environmental management approaches can also be applied to impede surface flows and enable water infiltration. These, however, are likely to be ineffective during extreme weather events.

Table 3. Options for responding to inland flooding.

Adaptation Options	References
Dykes or sluices (defend)	[30,35,41,44,49–54]
Creation or elevation of existing canal walls (defend)	[15,21,22,24,41,56,59]

- Infrastructure

Responding to flooding risk particularly focused on protecting local infrastructure that supports what is generally considered to be essential services are a subset of inland flooding. Identified in the literature are actions such as (1) securing infrastructure (e.g., by elevating roads and airports, protecting energy transmission lines and diversifying energy reticulation and sources); (2) increasing waste and water treatment capacity; and (3) reducing water flows that may cause inundation and necessitating construction of dykes, seawalls and elevated canal walls to prevent sea water inundation (Table 4). However, this last approach will offer no protection to flooding caused by freshwater run-off, which may occur due to river level rise or storm water run-off during heavy rain. In these cases, dykes and sea and canal walls may be maladaptive and exacerbate problems by creating barriers that impede water flow.

Table 4. Options for responding to the impacts of flooding on infrastructure.

Adaptation Option	References
Secure infrastructure	
Secure energy transmission lines (defend)	[6]
Elevate roads and airports (defend)	[60]
Redesign road system (defend)	[6]
Diversify energy supply (co-exist)	[6]
Increase waste and water treatment capacity	
Increase waste and water treatment capacity (co-exist)	[41]
Reduce water flows	
Drainage facilities and water pumps (defend)	[19,30,44,53]
Reduce paved areas to improve permeability of the soil or adopt water permeable pavements (defend)	[61]

• Housing

Reducing the risks associated with flooding to residential homes is a political issue for local authorities, with many community members assuming that local government will take responsibility for reducing risk. Adaptation options identified in the literature vary from being of short-term benefit and low cost (e.g., landscaping) to requiring costly structural modification of homes to significantly reduce risk from all but more extreme events (e.g., substitution of material and building techniques and codes so that houses are cooler, more resistant to flood and extreme weather events) (Table 5) The option of constructing dwelling levees might be appropriate in specific circumstances at the extreme of a predicted flooding area, but would probably attract neighbor criticism or result, in the flooding event, in an unflooded island isolated from essential services.

Table 5. Options for responding to the impacts of flooding on housing.

Adaptation Option	References
Levees around houses and other vulnerable structures (defend)	
Efficient drainage system (defend)	[62]
Gardens designed to safely redirect water (defend)	
Secure vulnerable equipment above the forecasted flooding level (co-exist)	
Design and use roofs capable of coping with high intensity rainfall events using impact and moisture resistant materials (e.g., metal rather than terracotta) (defend)	[6]
Design and reinforce existing structures (defend)	[20]
Maximize use of water-resistant materials (e.g., concrete, fiber cement) (defend)	[6]
Raise floor heights (defend)	[6,30,35,42]
Limited life of houses to minimize financial outlay (co-exist)	[6]
Multistory building with the lower level planned as non-living areas (co-exist)	[35]
Build as transportable and or floatable homes (co-exist)	[6]

(2) Fire

While bushfires are common in many parts of the world and are part of natural processes, longer drought periods are likely to intensify fires and increase their frequency. Hazard reduction burning of fire-prone areas to create a mosaic of patches (in Australia, from 1 to 7-year intervals between burns) is advisable but will need careful monitoring and adaptive management to ensure the desired mix of vegetation types. There is the potential for community debate on the benefits and efficacy of hazard reduction burning, so community engagement on the rationale, planning and implementation of a program is essential.

In 2019, the Sunshine Coast of Queensland, Australia experienced delayed arrival of the usual spring and summer storms and with strong winds and arson activity resulted in an ember storm never previously experienced. While loss of human life and property damage was relatively low because of effective and targeted fire control by emergency services, wildlife suffered and the vulnerability of homes became evident under extreme fire weather conditions that hazard reduction burning could not mitigate likely impacts. Risk reduction might involve creation of wider fire breaks around homes in a community that enjoys its leafy suburbs. In any case, the Sunshine Coast community will need to mitigate the impacts of fire with shutters and sprinkler systems in high-risk zones and using building materials that are fire resistant (Table 6). This will necessitate changes in building codes.

Table 6. Options for responding to the impacts of fire.

Adaptation Options	References
Shutters and sprinkler systems (defend)	[62]
Building materials that are fire resistant (defend)	[62]
Hazard reduction burns in fire-prone natural areas (defend)	[63]

3.2.3. Adaptation Options for Protecting Natural Systems

Major changes are expected to occur within land and marine natural systems in the next century due to not only climate change, but also deforestation, over-use of resources and pollution. Broad adaptation options to minimize disruptions to local natural systems are largely changed management that includes: (1) better environmental management, including the creation of artificial reefs and environments, (2) improved adaptive management of fire and efficient irrigation of natural and or restored areas; and (3) provision of incentives for conservation in farming areas, including benefits from carbon sequestration opportunities (Table 7).

Table 7. Options for responding to the impacts of climate change on natural systems.

Adaptation Options	References
Create artificial environments for the maintenance of species populations (defend) Provide incentive for conservation in farming areas, including benefits from carbon sequestration opportunities (defend)	[41]
Expand the protected area estate and revegetation (defend)	[6]
Establish ecological corridors (defend)	[41]
Translocate species at risk to secure locations (defend) Plan and plant gardens that provide habitat for native species and drought and flood resistant (defend) Improve composition of tree species in reforested areas (defend) Creation of artificial reefs (defend) Regulate the use of agritoxics (defend) Identify and protect climate change refuges (defend) Restoration of ecosystems (defend) Improve biodiversity management (co-exist) Efficient irrigation of natural and or restored areas (co-exist)	[19,62,63]
Prohibit or control the removal of beach sediment (co-exist)	[20]

3.2.4. Adaptation Options for Protecting Food Production

Due to change in mean temperature and rainfall, severe drought and extinction of animal pollinators, food production is predicted to be at risk and may prove to be one of the most significant threats to humanity from climate change [3]. Adaptation options proposed in the literature tend to be relatively low cost management actions that do, however, require strategic research to inform changed management practice: (1) adapt cropping techniques and species; (2) improve water, environmental and soil management; and (3) provide financial and technical assistance to farmers to deal with weather-related changes (Table 8).

Table 8. Options for responding to the impacts on farming and food production.

Adaptation Options	References
Adapt cropping techniques	
Adopt vertical farms (co-exist)	[64]
Substitute crops with drought and salt resistant cultivars (defend)	[20,41]
Plant an undercover to crops (co-exist)	
Diversify cropping species (co-exist)	[53–55]
Adjust planting and harvest dates (co-exist)	
Regulate the use of agritoxics that exterminate pollinizers (co-exist)	
Improve water, environmental and soil management	
Reforestation of areas likely to flood (defend)	[56]
Improve management to enrich the soil with organic matter (defend)	[56–61]
Improve irrigation systems and dig local dams (defend)	
Provide financial and technical assistance	
Financial and technical assistance to farmers (co-exist)	[56–61]

3.2.5. Adaptation Options for Protecting Availability of Fresh and Drinking Water

Water is one of the most important resources for life. Yet, the availability of fresh and drinking water are likely to be reduced by climate change. Historically, many regions of the world suffer intermittently from lack of water, and this situation is likely to be aggravated by climate change, not only because of drought periods, which are likely to last longer and be more severe, but also due to salinization of fresh water systems in coastal areas caused by sea level rise. Adaptation options identified (Table 9) are largely at the property level and include:

- Installation of devices to prevent seawater from back flowing into storm drains;
- Create farm dams and in other locations;
- Require households and businesses to install rainwater tanks to supplement the reticulated water supply system;
- Develop and apply desalinization technologies; and
- Create irrigation systems to ensure hydration of vegetation.

Table 9. Options for protecting the availability of fresh and drinking water.

Adaptation Options	References
Devices to prevent seawater from back flowing into storm drains (defend)	[62]
Dams in farms and in other different locations (defend)	[62]
Desalinization technologies (co-exist)	[20]
Household and business tanks to supplement the reticulated water supply system (co-exist)	[62]

3.2.6. Adaptation Options for Maintaining and or Improving Local Community Well-Being

Hotter weather and more intense storms are expected to affect community well-being. While emergency services will need to be prepared, households can act, at a cost, to reduce vulnerability. Adaptation options are largely housing improvements or modification (Table 10) but without guidelines and promoted smart practice, mandated building codes will probably be required that can be delayed by fear of political backlash to the household costs that would stem from retrospective regulation or increased cost of housing construction.

Table 10. Options for maintaining and or improving local community well-being.

Adaptation Options	References
Roofs capable of coping with high intensity rainfall events (defend) 'Green' roofs (defend) 'Green' infrastructure (defend) Improve natural ventilation of buildings (co-exist)	[6]
Better insulate homes (defend) Homes responding to sun orientation(defend) Hot air extraction technology (defend) Lighter-colored, reflective roofs (defend) Internal and box guttering material that can stand extreme weather conditions (defend) Double glazing of windows to support insulation (defend) Technology that decreases greenhouse gas emissions (defend) Well maintained roofs (co-exist)	[56]

4. Discussion

IPCC reports demonstrate that climate change is strongly linked to human activity, and 195 countries are committed to mitigation action to reduce greenhouse gas emissions as well as adapting its urban areas to respond to threats. All societies have a moral obligation to contribute, to the greatest extent possible, to reducing greenhouse gas emissions towards meeting international targets. For reducing greenhouse gas emissions, the primary driver for climate change mitigation lies within national and state energy policy and requires inter and intragovernmental cooperation for maximum efficacy. At the regional level, local authorities can show leadership by addressing their own emissions, informing businesses and households of mitigation actions they can voluntarily take, and by making bylaws that require mitigation action at the household and business levels. However, of immediate concern is the need to be prepared for climate change impacts by reducing risk and vulnerability of regional and household infrastructure: this requires encouraging and supporting adaptation.

4.1. Adaptation Spectrum

The options of defend, co-exist and retreat are not applied separately in urbanized areas. Instead, they progress as impacts become less manageable [30,33,34]. In the Torres Strait Islands (Australia), for example, seawalls were built as a defensive action to avoid flooding. As sea level rose, seawall efficacy diminished and flooding became more frequent [24]. Co-existence then took place until the situation became unmanageable and planning for retreat became an agenda item. In other words, while the academic literature presents the options of defend, co-exist and retreat as different adaptation approaches, they are stages within one spectrum of options. For this reason, in this work, the adaptation options presented were organized according to the component of interest in the affected system, with focus on properties and infrastructure, natural systems, food production, availability of water and well-being of the local population. The idea is that with the implementation of the identified options, retreat may be delayed or avoided.

4.2. Adaptation Options

While people may acknowledge the reality of climate change, they are less aware of its implications at a local and household level (partly because of the imprecision of modelling), and even less aware of actions they can take to contribute to reducing greenhouse gas emissions and how to adapt to reduce their vulnerability to impacts. In this context, local governments can play a significant role in improving community knowledge of climate change, its impacts, mitigation actions and adaptations that can be made in preparation for climatic and resulting change and extreme events. What will be important to communicate is that climate change mitigation action can be also action to reduce impacts. Engagement with local communities should be a pillar for meeting the obligations inherent in demonstrating commitment to sustainable development.

The 'soft' options for adaptation, such as awareness raising, planning, political articulation and financial incentives, insurance and professional skills enhancement, can address mitigation and adaptation concomitantly and be implemented immediately at relatively low cost. The likelihood of such efforts being wasteful is low because they align with sustainability and represent insurance against the worst of climate change impacts.

Community engagement is important because not all adaptation options can be implemented by governments who will be increasingly required to consider community-wide defensive actions in response to imminent threats to life and property. Residents and industries will need to modify their own properties and their management practices to ensure they are climate change ready, and not reliant on government to implement defensive measures as threats become imminent. Progressive implementation of adaptation strategies will have less upheaval of communities and individuals. Therefore, the responsibility of implementing adaption options in a timely manner at the regional level is shared between governments, residents and businesses. Yet, for adaptation to occur strategically and to optimize resource allocation, it should be orchestrated by local governments.

As climate change is likely to affect specific locations in multiple ways, including sea level rise, drought, extreme weather events and biodiversity extinction, it is unlikely that the implementation of just one adaptation option will address all climate change related problems. Hence, it is usually necessary to apply a systematic approach that addresses each threat and reduces the risk to affected areas and people. In this sense, further research would be necessary to develop a tool to assist in determining the most suitable and urgent adaptation options for each specific locale.

While the World Economic Forum has ranked extreme weather events as one of the most significant threats to humanity [2], it is important to keep in mind that risks are not only dependent on the hazard itself, but also on the strategies put on place to deal with challenges. The children's story *The Three Little Pigs* provides an analogy to better understand the risks that stem from weather events associated with climate change and the importance of adequate adaptation. In the story, while the same wolf menaces the three piglets, consequences significantly vary according to the structure protecting each of the piglets. With adequate warning, each piglet could better defend their existing infrastructure individually or collectively. The same is valid for the forecasted extreme weather events. While communities will be hit with the same perturbation strength, the consequences will vary depending on how each place, community and or person has prepared for the events. Those better adapted are likely to be less impacted.

5. Conclusions

While the scientific community largely agrees with the reality of anthropogenically-induced climate changes and the urgency to address it and its impacts, there remains uncertainty about how much temperature will change especially in specific locations. This is partly due to the complex ecological functioning of our planetary ecosystem and our limited understanding of the feedback processes involved in climate regulation, which compromises the development and the results of existing models. Yet, this is but part of the problem. The main challenge is to forecast how humanity will address the issue of reducing greenhouse gas emissions and how to respond to likely impacts in specific areas. Differences in forecasts of human response at national to local levels mean that the literature presents significant differences regarding predicted temperature rise and, consequently for example, by how much sea level is likely to rise, and appropriate response actions. IPCC works with conservative scenarios in which they consider temperature is likely to increase between 1.5 °C and 2 °C and sea level to rise between 0.26 m to 0.93 m by 2100, while the United Nations World Meteorological Organization estimates the temperature is more likely to increase between 3 °C to 5 °C by 2100 [1]. Under the best or worst scenario, adaptation remains necessary and urgent.

The importance of adaptation to environmental conditions as a mean to survival is not new; it was first defended by Charles Darwin in *Origin of Species* in 1859 [65]. Since then, it has been widely accepted that, not the strongest, nor the most intellectually capable, but the most flexible is more likely to survive. While Darwin's studies were not developed for the context of anthropogenic-induced climate change, his logic is equally valid for the forecasted warmer future. Therefore, paraphrasing Darwin, we can now say that those more likely to endure are those that are able to adapt and to adjust best to the changing environment in which they find themselves.

This work presented 80 adaptation options, mostly identified through systematic literature review. Not all options are adequate or necessary for all coastal communities; others would help mitigating impacts yet could be substituted by alternatives that would have a better cost-benefit in the specific region. The discussion here presented, focused on options that are generally suited for coastal areas. Suitability, in this case, relates to options that have positive cost benefits, with fewer negative impacts or with positive side effect consequences.

Adaptation actions can be high to low cost and appropriate at household/business to community levels. Local governments will be required to lead defensive adaptations that reduce risk community-wide. They are also obliged to inform their constituencies of defensive adaptations that allow for co-existence. In both cases, assessment of cost-efficacy in reducing vulnerability is necessary and will depend on evaluation of local conditions and the level of risk associated with threats.

Author Contributions: In developing this research article the authors by initials contributed the following: conceptualization, L.S. 40%, R.W.C. 60%; methodology, L.S. 40%, R.W.C. 60%; validation, L.S. 20%, R.W.C. 80%; investigation, L.S. 80%, R.W.C. 20%; data curation, L.S. 60%, R.W.C. 40%; writing—original draft preparation, L.S. 80%, R.W.C. 20%; writing—review and editing, L.S. 30%, R.W.C. 70%; visualization, L.S. 50%, R.W.C. 50%; project administration, L.S. 20%, R.W.C. 80%; and funding acquisition, R.W.C. 100%. All authors have read and agreed to the published version of the manuscript.

Funding: This research was funded by the Sunshine Coast Council and the University of the Sunshine Coast.

Conflicts of Interest: The authors declare no conflict of interest.

References

1. IPCC. Summary for Policymakers. In *Global Warming of 1.5 °C. An IPCC Special Report on the Impacts of Global Warming of 1.5 °C above Pre-Industrial Levels and Related Global Greenhouse Gas Emission Pathways, in the Context of Strengthening the Global Response to the Threat of Climate Change, Sustainable Development, and Efforts to Eradicate Poverty*; Masson-Delmotte, V., Zhai, P., Pörtner, H.O., Roberts, D., Skea, J., Shukla, P.R., Pirani, A., Moufouma-Okia, W., Péan, C., Pidcock, R., et al., Eds.; World Meteorological Organization: Geneva, Switzerland, 2018; p. 32.

2. World Economic Forum. *The Global Risks Report 2019*, 14th ed.; World Economic Forum: Geneva, Switzerland, 2019.

3. United Nations. Conference of the Parties: Adoption of the Paris Agreement. In *Framework Convention on Climate Change*; United Nations: Durban, South Africa, 2015.

4. Metz, B.; Davidson, O.R.; Bosch, P.R.; Dave, R.; Meyer, L.A. *Climate Change 2007: Mitigation of Climate Change*; Cambridge University Press: New York, NY, USA, 2007.

5. Boyd, R. Only Counting CO2, Not the Other Greenhouse Gases 2017. Available online: https://www.resilience.org/stories/2017-05-02/why-greenhouse-gas-emissions-did-not-really-stabilize-in-the-past-few-years/ (accessed on 4 March 2019).

6. Wardekker, J.A.; de Jong, A.; Knoop, J.M.; van der Sluijsac, J.P. Operationalising a resilience approach to adapting an urban delta to uncertain climate changes. *Technol. Forecast. Soc. Chang.* **2010**, *77*, 987–998. [CrossRef]

7. Unified Compliance Framework. n.d. Threat, in Compliance Dictionary. Available online: https://compliancedictionary.com/term/1614 (accessed on 30 April 2019).

8. IPCC. Summary for Policymakers. In *IPCC Special Report on the Ocean and Cryosphere in a Changing Climate*; Pörtner, H.O., Roberts, D.C., Masson-Delmotte, V., Zhai, P., Tignor, M., Poloczanska, E., Mintenbeck, K., Nicolai, M., Okem, A., Petzold, J., et al., Eds.; IPCC: Geneva, Switzerland, 2019.

9. Mills, M.; Leon, J.X.; Saunders, M.I.; Bell, J.; Liu, Y.; O'Mara, J.; Lovelock, C.E.; Mumby, P.J.; Phinn, S.; Possingham, H.P.; et al. Reconciling Development and Conservation under Coastal Squeeze from Rising Sea Level. *Conserv. Lett.* **2016**, *9*, 361–368. [CrossRef]

10. Bradley, M.; van Putten, I.; Sheaves, M. The pace and progress of adaptation: Marine climate change preparedness in Australia's coastal communities. *Mar. Policy* **2015**, *53*, 13–20. [CrossRef]

11. Colenbrander, D.; Bavinck, M. Exploring the role of bureaucracy in the production of coastal risks, City of Cape Town, South Africa. *Ocean Coast. Manag.* **2017**, *150*, 35–50. [CrossRef]

12. Gough, D.; Thomas, J.; Oliver, S. Clarifying differences between review designs and methods. *Syst. Rev.* **2012**, *1*, 28. [CrossRef]

13. Popay, J.; Roberts, H.; Sowden, A.; Petticrew, M.; Arai, L.; Rodgers, M.; Britten, N.; Roen, K.; Duffy, S. Guidance on the Conduct of Narrative Synthesis in Systematic Reviews: A Product from the ESRC Methods Programme Version. 2006, Volume 1, p. 92. Available online: http://citeseerx.ist.psu.edu/viewdoc/download?doi=10.1.1.178.3100&rep=rep1&type=pdf (accessed on 30 April 2019).

14. Berrang-Ford, L.; Pearce, T.; Ford, J. Systematic review approaches for climate change adaptation research. *Reg. Environ. Chang.* **2015**, *15*, 755–769. [CrossRef]

15. Obokata, R.; Veronis, L.; McLeman, R. Empirical research on international environmental migration: A systematic review. *Popul. Environ.* **2014**, *36*, 111–135. [CrossRef]

16. Serrao-Neumann, S.; Crick, F.; Harman, B.; Sano, M.; Sahin, O.; van Staden, R.; Schuch, G.; Baum, S.; Choy, D.L. Improving cross-sectoral climate change adaptation for coastal settlements: Insights from South East Queensland, Australia. *Reg. Environ. Chang.* **2014**, *14*, 489–500. [CrossRef]

17. Sidle, R.C.; Gallina, J.; Gomi, T. The continuum of chronic to episodic natural hazards: Implications and strategies for community and landscape planning. *Landsc. Urban Plan.* **2017**, *167*, 189–197. [CrossRef]

18. Department of the Environment and Energy of Australia & National Climate Change Adaptation Research Facility. 2017, Adaptation options for managing coastal risks under climate change Internet. Available online: https://coastadapt.com.au/adaptation-options (accessed on 5 February 2019).

19. Camare, M.H.; Lane, E.D. Adaptation analysis for environmental change in coastal communities. *Socio-Econ. Plan. Sci.* **2015**, *51*, 34–45. [CrossRef]

20. Hino, M.; Field, C.B.; Mach, K.J. Managed retreat as a response to natural hazard risk. *Nat. Clim. Chang.* **2017**, *7*, 364. [CrossRef]

21. Hino, M. Adapting to Climate Change through 'Managed Retreat'. 2017. Available online: https://www.carbonbrief.org/guest-post-adapting-climate-change-through-managed-retreat (accessed on 30 April 2019).

22. Knight, C. *Byron Shire Council—Coastal Hazard Planning Provisions*; NCCARF, Ed.; Snapshot for CoastAdapt, National Climate Change Adaptation Research Facility: Gold Coast, Australia, 2016.

23. Rainbird, J. *Adapting to Sea-Level Rise in the Torres Strait*; Case Study for CoastAdapt, National Climate Change Adaptation Research Facility: Gold Coast, Australia, 2016.

24. Douglas, E.M.; Kirshen, P.H.; Paolisso, M.; Watson, C.; Wiggin, J.; Enrici, A.; Ruth, M. Coastal flooding, climate change and environmental justice: Identifying obstacles and incentives for adaptation in two metropolitan Boston Massachusetts communities. *Mitig. Adapt. Strateg. Glob. Chang.* **2012**, *17*, 537–562. [CrossRef]

25. National Climate Change Adaptation Research Facility. *Cyclone Yasi—Communities Building Disaster Resilience. Snapshot for CoastAdapt*; National Climate Change Adaptation Research Facility: Gold Coast, Australia, 2016.

26. Ramm, T.D.; Watson, C.S.; White, C.J. Strategic adaptation pathway planning to manage sea-level rise and changing coastal flood risk. *Environ. Sci. Policy* **2018**, *87*, 92–101. [CrossRef]

27. Lin, B.B.; Capon, T.; Langston, A.; Taylor, B.; Wise, R.; Williams, R.; Lazarow, N. Adaptation Pathways in Coastal Case Studies: Lessons Learned and Future Directions. *Coast. Manag.* **2017**, *45*, 384–405. [CrossRef]

28. Buchori, I.; Pramitasari, A.; Sugiri, A.; Maryono, M.; Basuki, Y.; Sejati, A.W. Adaptation to coastal flooding and inundation: Mitigations and migration pattern in Semarang City, Indonesia. *Ocean Coast. Manag.* **2018**, *163*, 445–455. [CrossRef]

29. Fuchs, R.; Conran, M.; Louis, E. Climate Change and Asia's Coastal Urban Cities: Can they Meet the Challenge? *Environ. Urban. Asia* **2011**, *2*, 13–28. [CrossRef]

30. Minano, A. Visualizing flood risk, enabling participation and supporting climate change adaptation using the Geoweb: The case of coastal communities in Nova Scotia, Canada. *GeoJournal* **2018**, *83*, 413–425. [CrossRef]

31. Schernewski, G.; Schumacher, J.; Weisner, E.; Donges, L. A combined coastal protection, realignment and wetland restoration scheme in the southern Baltic: Planning process, public information and participation. *J. Coast. Conserv.* **2018**, *22*, 533–547. [CrossRef]

32. Martin, P.C.M.; Nunn, P.; Leon, J.; Tindale, N. Responding to multiple climate-linked stressors in a remote island context: The example of Yadua Island, Fiji. *Clim. Risk Manag.* **2018**, *21*, 7–15. [CrossRef]

33. Hurlimann, A.; Barnett, J.; Fincher, R.; Osbaldiston, N.; Mortreux, C.; Graham, S. Urban planning and sustainable adaptation to sea-level rise. *Landsc. Urban Plan.* **2014**, *126*, 84–93. [CrossRef]

34. Burley, J.G.; McAllister, R.R.J.; Collins, K.A.; Lovelock, C.E. Integration, synthesis and climate change adaptation: A narrative based on coastal wetlands at the regional scale. *Reg. Environ. Chang.* **2012**, *12*, 581–593. [CrossRef]

35. Lane, D.; Beigzadeh, S.; Moll, R. Adaptation Decision Support: An Application of System Dynamics Modeling in Coastal Communities. *Int. J. Disaster Risk Sci.* **2017**, *8*, 374–389. [CrossRef]

36. Lickley, M.J.; Lin, N.; Jacoby, H.D. Analysis of coastal protection under rising flood risk. *Clim. Risk Manag.* **2014**, *6*, 18–26. [CrossRef]

37. Smith, J.B.; Strzepek, K.M.; Cardini, J.; Castaneda, M.; Holland, J.; Quiroz, C.; Wigley, T.M.L.; Herrero, J.; Hearne, P.; Furlow, J. Coping with climate variability and climate change in La Ceiba, Honduras. *Clim. Chang.* **2011**, *108*, 457–470. [CrossRef]

38. Munaretto, S.; Vellinga, P.; Tobi, H. Flood Protection in Venice under Conditions of Sea-Level Rise: An Analysis of Institutional and Technical Measures. *Coast. Manag.* **2012**, *40*, 355–380. [CrossRef]

39. Taylor, B.M.; Harman, B.P.; Inman, M. Scaling-up, scaling-down, and scaling-out: Local planning strategies for sea-level rise in New South Wales, Australia. *Geogr. Res.* **2013**, *51*, 292–303. [CrossRef]

40. Griffith University. Storm Surge: Know Your Risk in Queensland! 2019. Available online: https://www.griffith.edu.au/__data/assets/pdf_file/0016/107314/CEMDSS-Storm-Surge-Community-Info-Sheet_Final.pdf (accessed on 6 February 2019).

41. Department of Environment and Heritage Protection. Coastal Management Plan. Queensland, 2013. Available online: https://www.qld.gov.au/__data/assets/pdf_file/0029/67961/coastal-management-plan.pdf (accessed on 5 February 2019).

42. Garner, G.G.; Keller, K. Using direct policy search to identify robust strategies in adapting to uncertain sea-level rise and storm surge. *Environ. Model. Softw.* **2018**, *107*, 96–104. [CrossRef]

43. Hanak, E.; Moreno, G. California coastal management with a changing climate. *Clim. Chang.* **2012**, *111*, 45–73. [CrossRef]

44. Hoshino, S.; Mikami, T.; Takagi, H.; Shibayama, T. Estimation of increase in storm surge damage due to climate change and sea level rise in the Greater Tokyo area. *Nat. Hazards* **2016**, *80*, 539–565. [CrossRef]

45. Mills, M.; Mutafoglu, K.; Adams, V.M.; Archibald, C.; Bell, J.; Leon, J.X. Perceived and projected flood risk and adaptation in coastal Southeast Queensland, Australia. *Clim. Chang.* **2016**, *136*, 523–537. [CrossRef]

46. Oh, S. Investment decision for coastal urban development projects considering the impact of climate change: Case study of the Great Garuda Project in Indonesia. *J. Clean. Prod.* **2018**, *178*, 507–514. [CrossRef]

47. Williams, S.J.; Ismail, N. Climate change, coastal vulnerability and the need for adaptation alternatives: Planning and Design examples from Egypt and the USA. *J. Mar. Sci. Eng.* **2015**, *3*, 591–606. [CrossRef]

48. Government of Queensland, Department of Environment and Resource Management, Brisbane. Queensland Coastal Plan. 2011. Available online: https://www.qld.gov.au/__data/assets/pdf_file/0029/67961/coastal-management-plan.pdf (accessed on 1 April 2019).

49. Chow, J. Mangrove management for climate change adaptation and sustainable development in coastal zones. *J. Sustain. For.* **2018**, *37*, 139–156. [CrossRef]

50. Wang, C.H.; Baynes, T.; McFallan, S.; West, J.; Khoo, Y.B.; Wang, X.; Quezada, G.; Mazouz, S.; Herr, A.; Beaty, R.M.; et al. Rising tides: Adaptation policy alternatives for coastal residential buildings in Australia. *Struct. Infrastruct. Eng.* **2016**, *12*, 463–476. [CrossRef]

51. Lovett, J.; Useche, D.C.; Rendeiro, J.; Kalka, M.; Bradshaw, C.J.A.; Sloan, S.P.; Laurance, S.G.; Campbell, M.; Abernethy, K.; Alvarez, P.; et al. Averting biodiversity collapse in tropical forest protected areas. *Nature* **2012**, *489*, 290.

52. Hallegatte, S. Strategies to adapt to an uncertain climate change. *Glob. Environ. Chang.* **2009**, *19*, 240–247. [CrossRef]

53. Zeppel, H. Local planning for climate adaptation in coastal Queensland. In Proceedings of the 4th Queensland Coastal Conference 2013, Townsville, Australia, 2–4 October 2013.

54. Bush, D.M.; Neal, W.J.; Young, R.S.; Pilkey, O.H. Utilization of geoindicators for rapid assessment of coastal-hazard risk and mitigation. *Ocean Coast. Manag.* **1999**, *42*, 647–670. [CrossRef]

55. Drake, J.A.; Bradford, A.; Marsalek, J. Review of environmental performance of permeable pavement systems: State of the knowledge. *Water Qual. Res. J.* **2013**, *48*, 203–222. [CrossRef]

56. Department of Climate Change and Energy Efficiency. Housing. Adapting to Climate Change. 2013. Available online: http://www.yourhome.gov.au/sites/prod.yourhome.gov.au/files/pdf/YOURHOME-Housing-AdaptingToClimateChange.pdf (accessed on 5 February 2019).

57. Loehle, C. Applying landscape principles to fire hazard reduction. *For. Ecol. Manag.* **2004**, *198*, 261–267. [CrossRef]

58. Department of Water and Environmental Regulation, Government of Western Australia. Ecosystems and Biodiversity. Adapting to Climate Change 2019. Available online: https://www.der.wa.gov.au/your-environment/climate-change/254-adapting-to-climate-change?showall=&start=5 (accessed on 6 February 2019).

59. Fletcher, C.S.; Taylor, B.M.; Rambaldi, A.N.; Harman, B.P.; Heyenga, S.; Ganegodage, K.R.; Lipkin, F.; McAllister, R.R.J. *Costs and Coasts: An Empirical Assessment of Physical and Institutional Climate Adaptation Pathways, in CSIRO Climate Adaptation Flagship*; National Climate Change Adaptation Research Facility; CSIRO: Southport, Australia, 2013; p. 62.

60. Hsu, J. Sink or Swim: 6 Ways to Adapt to Climate Change. 2012. Available online: https://www.livescience.com/22210-adapt-survive-climate-change.html (accessed on 6 February 2019).

61. United Nations Climate Change. What Do Adaptation to Climate Change and Climate Resilience Mean? 2019. Available online: https://unfccc.int/topics/adaptation-and-resilience/the-big-picture/what-do-adaptation-to-climate-change-and-climate-resilience-mean (accessed on 6 February 2019).

62. Global Agriculture. Agriculture at a Crossroads Internet. 2019. Available online: https://www.globalagriculture.org/report-topics/adaptation-to-climate-change.html (accessed on 6 February 2019).

63. Dooley, E.; Frelih-Larsen, A. Agriculture and Climate Change in the EU: An Overview. Climate Policy Info Hub. 2015. Available online: https://climatepolicyinfohub.eu/agriculture-and-climate-change-eu-overview (accessed on 6 February 2019).

64. Lloyd, M. *UN Warns World on Track to Breach 3C Rise by 2100*; Last Year was Fourth Warmest on Record; ABC News: New York, NY, USA, 2019.

65. Darwin, C. *On the Origin of Species, 1859*; Routledge: London, UK, 2004.

 *climate*

Article

Perception of Environmental Spillovers Across Scale in Climate Change Adaptation Planning: The Case of Small-Scale Farmers' Irrigation Strategies, Kenya

Volenzo Tom Elijah * and John O. Odiyo

School of Environmental Sciences, Department of Hydrology and Water Resources, University of Venda, Thohoyandou 0950, South Africa; john.odiyo@univen.ac.za
* Correspondence: volenztom@gmail.com; Tel.: +27-73-4561800

Received: 24 October 2019; Accepted: 9 December 2019; Published: 26 December 2019

Abstract: The failure to acknowledge and account for environmental externalities or spillovers in climate change adaptation policy, advocacy, and programming spaces exacerbate the risk of ecological degradation, and more so, the degradation of land. The use of unsuitable water sources for irrigation may increase salinisation risks. However, few if any policy assessments and research efforts have been directed at investigating how farmer perceptions mediate spillovers from the ubiquitous irrigation adaptation strategy. In this study, the cognitive failure and/or bias construct is examined and proposed as an analytical lens in research, policy, and learning and the convergence of disaster risk reduction and climate change adaptation discourses. A cross-sectional survey design and multistage stratified sampling were used to collect data from 69 households. To elicit the environmental impacts of irrigation practices, topsoil and subsoils from irrigated and non-irrigated sites were sampled and analysed using AAS (atomic absorption spectrophotometer). A generalised linear logistic weight estimation procedure was used to analyse the perception of risks while an analysis of variance (ANOVA) was used to analyse changes in exchangeable sodium percentage (ESP). The findings from small-scale farmers in Machakos and Kakamega counties, Kenya, suggest multifaceted biases and failures about the existence and importance of externalities in adaptation planning discourses. Among other dimensions, a cognitive failure which encompasses fragmented approaches among institutions for use and management of resources, inadequate policy. and information support, as well as the poor integration of actors in adaptation planning accounts for adaptation failure. The failures in such human–environment system interactions have the potential to exacerbate the existing vulnerability of farmer production systems in the long run. The findings further suggest that in absence of risk message information dissemination, education level, farming experience, and information accumulation, as integral elements to human capital, do not seem to have a significant effect on behaviour concerning the mitigation of environmental spillovers. Implicitly, reversing the inherent adaptation failures calls for system approaches that enhance coordinated adaptation planning, prioritise the proactive mitigation of slow-onset disaster risks, and broadens decision support systems such as risk information dissemination integration, into the existing adaptation policy discourses and practice.

Keywords: adaptation failure; adaptation planning; economic interests; climate change; ecosystem spillovers; policy; risk perception; transformation

---

1. Introduction

Though climate change is used as justification for environmental and livelihood interventions [1], there is a risk of adaptation failure or an inability of adaptation action to meet set objectives and/or generate hybrid risks, such as environmental degradation [2,3]. Accordingly, disaster risk drivers such as poor land management, unsustainable use of natural resources, and declining ecosystems have emerged as focal points in climate change action and the pursuit of sustainable development goals [4,5]. The growing evidence of links between climate change adaptation (CCA) and disaster risks has also seen concomitant efforts at integrating disaster risk reduction (DRR) and CCA [6], with a focus on the dialectical and/or trialectic tension between resilience, adaptation and risk management within the broader social-ecological system approach, particularly the human-environment nexus [3,7]. Analytical lenses that link climate change adaptation to other drivers of change has thus emerged as essential for effective adjustment to changing climate stimuli [8].

Comprehensive adaptation planning frameworks address policy and implementation process interlinkages or scales at local and national levels [3,9,10]. Implicitly it encompasses the integration of sustainable development and disaster risk management lenses [11–13], policy engagement or framing [2,3,9,14], as well as changes in policies and institutional arrangements that mediate successful scaling up of CCA [1]. Risk management and robust decision making are core features that address underlying risks [15], more so responses to adaptation needs that span long time horizon [16]. Focusing on implementation phase in adaptation planning is critical as statements of intent, allocated resources, and envisioned alternatives in the form of programs, legislation and rules on their own cannot guarantee effective solutions to collective adaptation needs [17,18].

Innovative lenses on deliberations about risk appraisal [18], the role of values, interests, and institutions that constrain the societal response to change and unpacking of underlying causes are some of the factors of interest in the emerging approaches to climate risk management [11,14]. However, in spite of the recognition of the need to integrate DRR, climate change, and sustainable development, and their successes at the conceptual level, insufficient interrogation of the underlying risks tend to bias disparate adaptation planning discourses towards business as usual (BAU) implementation trajectories that undermine the effectiveness of adaptation action [19,20]. Most importantly, BAU or routine adjustment to adverse impacts from climate change tend to ignore social costs which are at cross purpose with some of the tenets of sustainable development. There is an urgent need, therefore, to reorient adaptation planning frameworks to minimise the risk of adaptation failure.

Social structures mediate the exchange of knowledge and behaviour, such as the development and diffusion of adaptation technology to climate change [21]. Cognition or knowledge about risks and shared understanding could build coherence and vision into integrative frameworks, such as those that concurrently address sustainability and disaster risk reduction [11,21]. Accordingly, values, beliefs, interests, knowledge and expectations are considered integral to holistic approaches and effective adaptation [3]. However, many of the existing integrative models are constrained as they fail to recognise the centrality of individuals [11]. Additionally, current integrative models pay little attention to time-related concerns that may amplify the risk of slow-onset disasters [22].

The individual agency and wider pathways of change which portend challenges in adaptation discourses [23], are related to the complex social networks and relations in which people are embedded, commitments and understanding of social and ecological risks [7,14]. Accordingly, complementary efforts that address questions of scale, fit, and interplay in policy and governance could partly resolve such dilemmas [24,25]. In this article, we explore how multifaceted biases and failures with respect to the existence and importance of negative externalities constrain system integration in adaptation planning discourses.

Though integration of CCA and mitigation of associated disaster risks or ecosystem spillovers, such as salinisation risks, can be advanced through theoretical and/or conceptual multiplicity [26], convergence of CCA and DRR is constrained in agricultural production systems [7]. The constraints are related to difficulties in the integration of learning, reflectivity, and change management, as well as, lack of institutionalisation of CCA-DRR into the planning process [11,14]. More specifically, there is a paucity of knowledge in diagnostic procedures and empirical evidence that illustrate conceptual and theoretical convergence, as well as urgency for action [2]. Specifically, there are gaps in adaptation policy framing regarding potential mechanisms for the integration of CCA-DRR models [6]. We posit that environmental externalities have great potential to facilitate a holistic vision for the convergence and operationalisation of the often disparate CCA-DRR approaches.

Though system integration at local and global scales in sustainability discourses have emerged [7], there is still little attention paid to environmental spillovers [27,28]. Such limited attention to environmental spillover effects is more widespread in climate change action. In risk analysis, fast and frugal heuristics is adopted if ignoring some information does not compromise accuracy of the findings [29]. We adopt the logic and concur with Reed et al. [30] and Reid and Coleen [31] that thresholds and sustainability indicators on a limited number of parameters, such as soil health (including qualitative aspects, such as salinity levels), could be used as empirical indicators to assess the effectiveness and/or failure of adaptation strategies, such as irrigation. In particular, we adapt [32] in that temporal variation in soil salinity is an appropriate indicator in the monitoring of degradation risks and proxy for sustainability trends.

To illustrate our proposition, we assess various dimensions of cognitive failures and/or biases in autonomous adaptation pathways and how this constrain transformative adaptation discourses among small-scale farmers. Building upon the above assumptions, we employ a survey study and assessment of salinity dynamics to unpack the interplay between cognitive failure, environmental externalities and adaptation failure. The quantified changes and significance interpretation is based on FAO [33] classification of salinity risks from irrigation water.

By unpacking the poorly understood environmental spillover effects, we provide insights that complement and enhance the utility of existing transformative adaptation planning frameworks. The nested adaptation assessment model thus provides holistic lenses that address multifaceted biases at policy, research and implementation levels. The model addresses complex interplay between the climate system, the human system, as well as sustainability concerns, related policy analyses and ultimately system integration in adaptation planning. In so doing, the study contributes to the development of a robust and innovative diagnostic approach that integrates empirical data, cognitive and scale dynamics (such as, institutional polices, farmer management practices) in projecting adaptation failure.

2. The Multifaceted Dimensions to Cognitive Construct In Adaptation Policy

The multifaceted dimensions to cognitive failure and/or bias construct in adaptation planning discourses is presented hereunder.

2.1. The Policy–Practice Divide as Cognitive Failure

The development paths and the choices that define adaptation choices have greater bearing on the severity of future climate impacts, local-scale disaster risk reduction (DRR) and resource management [34], as well as broader social dimensions, such as risk perception [35]. Though planned adaptation presents new opportunities in the mitigation of climate change related risks [36], reactive or autonomous adjustments to adverse climate stimuli and the associated investments may increase the risk of maladaptation, hence an increased exposure of ecosystems, sectors, or social groups to hybrid or secondary risk [19,37,38]. For example, the adoption of technologies in water management such as in flood control, has potential for new downstream hazards, in itself an example of negative interactive impacts between adaptation, governance failures and disasters [39]. The environmental damage and lack of fit for purpose associated with such interactions has been termed as adaptation failure [2,3,9].

Optimising the benefits and concomitant minimisation of maladaptation risks through robust adaptation, mitigation, and sustainability frameworks has emerged and been suggested as a triple win strategy in adaptation policy framing [3,9,40,41]. Accordingly, the effective formulation of adaptation strategies, as well as the success of CCA policy and programming in climate risk management, to a large extent, is predicated on local knowledge of adaptation [42], local context of adaptation strategies [43,44], as well as agent perception [9,45,46]. In addition, effective adaptation depends on policy support that facilitates environmental sustainability, as well livelihood capital, such as financial returns and knowledge stocks [43]. The identification of causes, agents, and flows behind the externalities or spillovers is thus critical to understanding mitigation of externalities [7,24].

Decision making is unpacked through adaptation activity and solution spaces, such as individual, technology, livelihoods, behaviour, the environment, institutions, popular, and policy discourses [1]. Enhancing better understanding and managing effects across multiple systems and scales is thus critical in sustainability policy and management. In particular, the use of human perception lenses has immense potential in promoting system resilience [7,47]. However, individual adaptation hinges on whether an impact, anticipated or experienced, is perceived as a risk and whether it should and/or is acted upon through adaptation policies, or is constrained by inertia and cultures of risk denial [21]. This necessitates the use of holistic approaches that consider feedback loops to shape outcomes from the complex interplay between the climate system, the human system and ecosystems, as well as an assessment of sustainability [2,7,9,48].

The multiple interactions between governance and resource users' systems are consequential on provision of ecosystem goods and services, as well as, externalities [24]. Accordingly, under the sustainable development paradigm, ecological considerations are prioritised over short-term economic pay-offs [49]. In situations of inadequate information, and where alternatives and consequences are not well understood, the polluter pay and the precautionary principle [50] are widely accepted as complimentary to legislative and enforcement mechanisms [50,51]. However, for most developing nations, the precautionary and polluter pay principle have been adjudged to be ineffective in the mitigation of environmental externalities [2,52]. The pursuit of sustainability has thus been re-oriented to encompass coordination mechanisms and integrative use of social ecological lenses that unpack the complex interplays between agent cognition, governance, social and policy discourses with regard to outcomes, such as environmental externalities [7,24]. In essence, synergies and trade-offs between broader development goals and climate-risk management are the focus in adaptation planning [2,53]. However, environmental spillovers or downstream costs, such as salinisation, have received little attention in such discourses.

Though agent behaviour across scale, the processes in behaviour development, as well as behaviour patterns can be exploited in scenario building of likely spillover impacts [54], there is lack of understanding and concern for important linkages between natural resource management, development, DRR and climate change mitigation and adaptation constrain systemised planning [19,55]. For instance, policy makers, depending upon their institutional mandates, may view a single hazard, such as waterborne diseases and flooding, separately, instead of multiple, interrelated hazards at one time [9,39], as well as demonstrate a bias to immediate adaptation needs during policy framing and decision making [14].

Reducing the risk of adaptation failure depends on the extent to which multiple actors across scale and the broader social contexts are integrated into decision making [2,14,19,56,57], as well as responsive legislative frameworks [56]. Information and policy coherence [9] as well as the coordinated framing of the problem among actors with influence on adaptation planning and policy tend to substantially reduce such risks [19,57]. Policy and information support frameworks have great potential to guide informed decision making and a paradigm shift towards effective adaptation action in general, as well as learning about, and mitigation of negative social and environmental externalities in particular [9].

Though adaptation-mitigation-sustainability frameworks exist, accounting for environmental spillovers in planning processes remains as a challenge [7]. Such a challenge is routinely encountered in search of solutions to environmental change problems with intractable feedback loops [58]. By default, the favoured technology end state solution approaches in routine adaptation discourses fail to acknowledge and account for environmental footprints [59,60]. As adaptation and mitigation in agriculture are country and farmer specific and by farmer characteristics, such as farm size and education level [61], risk reduction planning process involves a diverse solution space, such as, knowledge of situations (cognition), processes and systems [3,5,11,14]. The low institutional awareness and institutional coordination between agencies responsible for disaster management and climate change adaptation, as well as overall development planning thus tend to entrench the reactive and/or fragmented adaptation solutions [6,12].The divergence is reflective of cultural cognitive institutions that affect system understanding, boundary setting and participatory search for solutions [2,11]. This may result into biased planning frameworks and adaptation failure [17,39,62]. Implicitly holistic approaches that pay attention to feedback loops between the climate system and the human system are invaluable in adaptation planning [48]. In particular, multi-hazard and multisectoral frameworks that foster people centered, collaborative partnerships, mechanisms and institutions for implementation of instruments relevant to building resilient socio-ecological systems are critical.

2.2. Cognitive Failure and Mitigation of Ecosystem Risks

Though the three domains of adaptation, mitigation and productivity are dialectically related to the other two and thus intricately intertwined [63], operationalising system convergence is undermined by absence of over-arching national policies that integrate CCA and DRR into various aspects of land-use planning and typified by lack of capacity to assess, interpret and apply data on climate change risks and vulnerabilities. Convergence is also undermined by bottlenecks in the integration of plans among and within agencies [12]. The dissonance between individual values and formalised institutions and organisations as entry points for alternative adaptation pathways [23], and convergence between CCA and DRR is thus likely to demand substantial institutional changes [6].

Knowledge of consequences, their causes, and implications play a role in peoples risk belief and mitigation actions [64]. Cognition or perception aid in mobilising peoples' commitment to action over environmental problems [65]. Perception of risk, habit, social status, and age as individual attributes are thus critical in collective action decision-making [21]. At the community level, analytical and conceptual lenses that unbundle cognitive biases and failures, as well as integrate and transform individual and collective agency, are critical to risk reduction and resilience building [66]. Theoretical and empirical multiplicity lenses improve analytical rigour, address conceptual and knowledge gaps, and solve complex problems and contextual dilemmas while encouraging synergies [26,58]. The utility of communication in CCA-DRR convergence discourses at different instutional scales [6,67], as well as development and dissemination of adaptation technology options [68], is thus critical.

The increase in risk and vulnerability from climate extremes calls for increased attention to an array of underlying drivers and lenses, such as, ecosystem services, governance and information needs [24]. However the dilemma arises due to divergence in priorities at different times and scales hence the need for analytical and policy innovations that advance and/or broker complementarity in CCA policy, advocacy and programming spaces [1,67]. However, the complex human–environment system feedbacks are potential dilemmas that may constrain planning. For example, though awareness plays a critical role in disaster mitigation [64], increased information may be ineffective as a tool for better decision making where profit motive (proxy for risk disposition) prevails [69]. Intuitively there is a need for innovative lenses that resolve inherent value conflicts around immediate private gain and long-term social concerns.

Though changes in external stimuli, such as temperature and moisture are sources of risks that trigger development of robust adaptation strategies at micro i.e., individual farm level [70], the farmer as a primary actor in adaptation planning, is motivated by short-term reactive incremental adaptation preferences that are biased towards immediate economic interests and/or survival objectives other than long-term sustainable risk reduction initiatives [9,46,71]. The prioritisation of narrow economic interests and immediate payoffs as opposed to long term social good, discounts the importance of future risks and undermine sustainability of ecosystems [1,72].

Though collective action and public support is a necessary condition for the effectiveness of mitigatory action (i.e., internalisation of environmental effects, such as methane emissions, salinity spillovers etc.), the accruing benefits from such action, are felt after long time lags and spread or diffused to the wider social system. The extra costs in internalising the spillovers reduces incentives for individual actor action [46,73]. The rationale seems to account for popularity of adaptation pathways that do not address negative ecosystem externalities or spillovers. In essence, effective adaptation planning, moreover the mitigation of slow onset disaster risks, should consider the integration of short term and long-term social interests.

In climate change adaptation, sustainability is often framed as a one way driver of change in the system of interest with little attention to feedbacks between the system of interest and other systems [7,19,74,75], as well as poor cognition of spillover systems [76]. The cognitive barriers are linked to poor quality and/or lack of specific information, poor coordination across scale [9], fragmented understanding among the actors [2,77], as well as operational challenges among constrained agents [3]. Cognitive failure and/barriers thus inhibit informed and sustained action [78]. The failure is exacerbated by ineffective implementation and/or poor enforcement mechanisms [2,9], especially the mismatch between expert and lay perceptions of risk [79]. More importantly, most policy framings in CCA fail to consider externalities by favouring short term political needs [14,39,80].

The bias towards immediate payoffs across scale increases the need for integration and use of perception at community level in the design, analysis and policy reframing on adaptation planning [1,11,14]. Dissemination of information on such risks or risk communication, has been found to play a critical role in the abatement of externalities [81]. The framing of communication regarding the mitigation of future risks is thus critical as it affects cognition and disaster risk reduction responses [65,82]. In particular, variation in perception is an important consideration because differences between lay and expert perceptions of risk impact the success of risk communication [79]. Investigating farmer perceptions could provide novel insights and advances in the concomitant integration of sustainability, disaster risk reduction, resilience building and development planning lenses into transformative adaptation discourses, the identification of governance gaps and betterment of system integration frameworks.

2.3. Underlying Risks and Transformative Adaptation

The extent to which underlying risks are addressed defines whether the adaptation pathway is transformative or incremental. Several pathways such as transformation, vulnerability reduction, disaster prevention, preparedness, response and recovery, and building resilience provide solution spaces for risk management and adaptation to extreme climate changes [83]. While incremental adaptation relies on BAU trajectories, transformative adaptation considers alternative development priorities, preferences and pathways that address the social drivers and processes. It thus incorporates early warning systems as disaster risk reduction tools and lens into planning processes [1,2,9,14]. Implicitly, transformative adaptation includes monitoring, evaluation and learning for improvement and policy support [9]. Operationalising transformative adaptation has however received less attention in practice [14,81].

Incremental adaptation discourses primarily focus on technical approaches to improve predictive capabilities in adaptation planning cycle [2,9,14]. Incremental adaptation frameworks are thus short of social lenses that can unpack underlying risks. In contrast, transformative adaptation frameworks address deep rooted causes of risk and vulnerability with the primary objective being to enhance co-benefits and minimise the risk of the adaptation deficit or failure [14,84]. Enabling drivers towards transformative discourses include the upstream dialogue and exploration of values and visions about future decision making processes [85]. Increased awareness on the less acknowledged salinisation risks could aid such forward looking planning.

The scaling up of adaptation could provide multiple co-benefits where public participation, awareness raising campaigns, law enforcement, as well as strong political exist [86]. Improved access to information about appropriate adaptation strategies appear to support adaptation processes and resilience building at the local level [11,44], as well as raise procedural questions for decision-makers [1]. Accordingly, engagement with individuals might be a useful lens through which communities and practitioners are sensitised to risks with a positive impact on the construction of a more dialectical approach to DRM/CCA and sustainable development [14]. We argue that transformation pathways should revolve around the multifaceted cognitive failure construct and environmental externalities.

Though media can be exploited to enhance the understanding of disasters, especially where vicarious experience is concerned [87], some authors [88] have found no relationship between exposure to sources of information or self-rated knowledge about climate change and support for climate change policy. Such a dilemma could be resolved partly through participatory communication [89] and the concomitant use of seamless support systems, such as, risk communication, which have great potential to address cognitive biases and/or failures [81].

2.4. Salinity Footprints and Adaptation Failure

Water quality and its suitability for use in irrigation is judged on potential severity of problems that can be expected to develop during its long term use [33,90]. The total concentration of soluble salts (salinity hazard) in terms of electro-conductivity (EC), relative proportion of sodium to other principal cations (sodium hazard) expressed as sodium adsorption ratio (SAR), bicarbonate concentration relative to the concentration of calcium plus magnesium and boron hazards, or the concentration of boron or other toxic elements are the most important determinants of quality and suitability of water for irrigation [90].

Salinity is recognised as one of the greatest land degradation processes and declines in soil productivity, especially in arid and semi-arid regions [91,92]. High levels of salts in water used for irrigation has been implicated to affect soil fertility and crop yield [93]. Salinity hazards or EC exceeding certain threshold levels reduce water availability in the root zone and cause 8–86% drop in crop yields [33]. Such risks increase with use of ground water (e.g., from boreholes) of high salt content for irrigation [94]. In particular, salinity negatively alters soil microbial and biochemical properties, metabolic efficiency and growth of soil microbes [95]. Though salinity in soils tend to vary significantly, it indirectly impacts climate change through oxide (N_2O) emissions, and hence has an effect on global warming [96].

While primary salinisation is associated with parent material mineralogy, secondary salinisation is dependent on agronomic practices, such as fertilization, poor drainage and use of inappropriate water sources [32,97]. In a study of groundwater quality in the Soutpansberg fractured aquifers, South Africa, agricultural activities produced localised impacts in terms of elevated concentrations of calcium, chloride, magnesium and nitrates in groundwater [98]. Where small scale production systems dominate, the underestimation of cumulative impacts of the seemingly minor individual footprints may result in an ecological disaster in the long run.

Land degradation is one of the slow onset disasters with adverse social and ecological impacts [99]. For example, in India, one of the countries where land degradation is widespread, six million hectares of the 147 million hectares of land classified as degraded is attributed to salinisation [100]. Though slow-onset disasters, such as land degradation generally do not result in sudden fatalities or casualties and acute property damage, they are more extensive in their impact and more destructive in the long term than rapid-onset disasters such as floods, hurricanes, and earthquakes [101]. Since individuals may not recognize land degradation as an underlying cause of vulnerability, awareness of such a type of a disaster is critical [102]. A lack of and/or poor knowledge of the consequences of the effect of such slow-onset disasters, such as those associated with spillovers from salinisation, fits the narrative of adaptation failure and demonstrates the intractable challenges between adaptation action and vulnerability to induced risks or spillover effects.

2.5. The Agricultural Sector, Climate Change Risk and Adaptation Policy Context In Kenya

Kenya is predominantly an agrobased economy where small scale farmers dominate with about 75% of the populations' livelihoods directly linked to agriculture [103]. Agriculture is thus key to overall national development, equity objectives and sustainable growth. Intuitively, weather-related disasters, particularly droughts, present a major challenge to the predominant rainfed agricultural production system with profound adverse impact on the economy. The adverse effects negatively affect foreign exchange earnings, food security and nutrition, employment and rural livelihoods. Adaptation to extreme weather impacts is thus a priority under National Adaptation Policy Action plans (NAPAs). Among other objectives, NAPAs envisages improved crop productivity through irrigation [104].

Adaptation to climate-related risks is expected to be achieved within a number of institutional and governance frameworks, such as the climate change Act and the Environmental Management Coordination Act (EMCA) which directly or indirectly impinges on agricultural sector planning. EMCA is a framework legislation under the stewardship of the National Environment Management Authority (NEMA), the government agency for coordination, enforcement and compliance on all matters on environment. As the principle instrument that establishes the legal and institutional framework for all matters that touches on environmental management in Kenya [105], EMCA adopts the "precautionary principle" as a sustainability safeguard in decision making. The 1st Schedule of the EMCA act, parts vi and vii provides for the process and projects that should undertake environmental impact assessments (EIA), audit (EA), and monitoring respectively. Irrigation is among projects that should undertake EIA/EA. However, the act only refers to effluents and not the processes nor the slow onset disaster risks, such as salinisation.

Building farmer resilience to climate change risks is the main objective under the Agricultural Sector Transformation and Growth Strategy [103], which in agriculture operationalises the climate change act. Though the Climate Change Act [106], broadly addresses mechanisms and measures towards low carbon climate development, it fails to address environmental externalities, such as salinity footprints, an ubiquitous adaptation pathway in the country. The Agricultural Sector Transformation and Growth Strategy envisages an increase in access to irrigation by small scale farmers from the current level of 5% to 11%.

3. Methodology

3.1. Study Area

The location of study sites, Likuyani subcounty in Kakamega County and Mavoko subcounty in Machakos county respectively is provided in Figure 1. Though the study sites are located in contrasting ecological zones, both are highly populated and characterised by high poverty levels. High population and poverty levels are drivers of increased livelihood vulnerability to climate change related risks. Kakamega covers an area of 3051 km^2 with a population of 1,660,651(approximated growth rate of 2%), that translates to population density of 544.3/Km^2. Machakos covers an area of 6208 km^2 with a population of 1,098,584 persons (projected growth rate approximated at 1%), and a density of 177.0/Km^2 [107].

Kakamega county is located in Western Kenya between longitude 34^0 35^1 E and latitude 0^0 and $0^0 15^1$ N [108]. The county is characterized by commercial sugarcane farming as well as maize production at subsistence and commercial level as major economic activities [107]. Agriculture employs 80% of the population and is critical to poverty (currently at about 50%) reduction in the county [109]. The Agro ecological zones (AEZs) range from UM1 (upper middle-1) to LM-3 (lower middle-3) hence variation in rainfall, agricultural potential and productivity in terms of livestock type, crop varieties and actual/potential yield levels [108]. Most of the soils in the county are thus heavily leached due to high rainfall and relay cropping. An agro-ecological zone describes agronomic conditions on basis of landform, soil types, rainfall, temperature and water availability, which in turn influences the type vegetation, length of crop growing period and their adaptability to the locality [110]. The county receives 1200–2200 mm of rainfall per annum with the first rains of 500–1100 mm and second rains of 450–850 mm. However, farmers in the area, notably the northern part (the study site), is affected by extreme climate change extremes in form of droughts. The extreme weather episodes are exacerbated by high evapo-transpiration that averages 1600 to 1800 mm. Generally, the county has experienced warming trends, interannual variability in the amounts of rainfall evidenced through increased number of consecutive dry days, as well as intense downpours that occasion flooding [109].

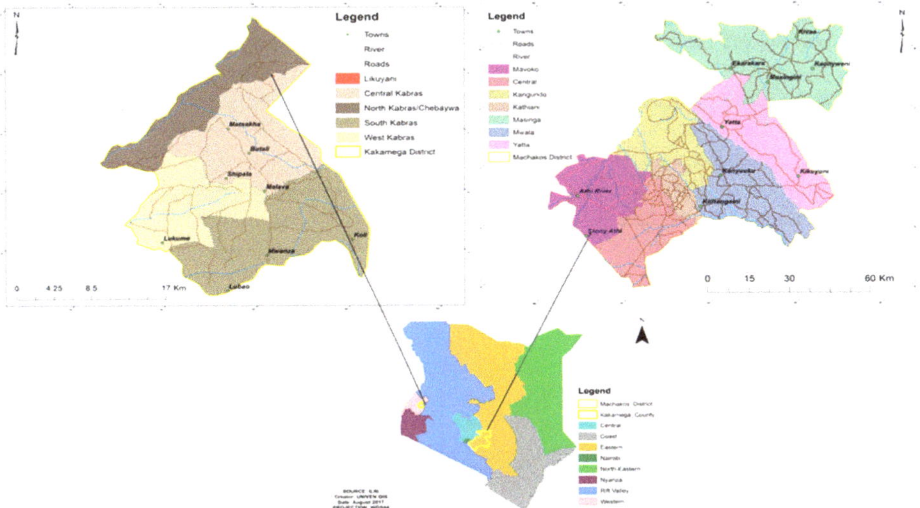

Figure 1. Geographical Information System (GIS) Generated map of study sites in Kakamega and Machakos Counties, Kenya.

Machakos county is located in Eastern Kenya, between latitudes 0°45′and 1°31′S and longitudes 36°45′ and 37°45′E and an elevation of between 790 and 1594 m above sea level. The agriculture economy in the county contributes 70% of household income and is characterized by livestock farming, as well as small-scale crop production at subsistence and commercial levels [107]. The AEZ range from LM2 (lower middle-1) to LM-3 (lower middle-3). The county is characterised by a semi-arid type of climate (except in highland areas) and cool to hot temperatures that averages 18 °C and 29 °C. It receives bimodal but unevenly distributed and unreliable rainfall that averages 500 mm to 1300 mm annually. The agricultural potential and productivity in terms of livestock type, crop varieties, and actual/potential yield levels is thus highly limited by the low moisture potentials. This increases vulnerability of farmers to production failures. The absolute poverty in the county averages about 61% [111].

3.2. Data Collection

For this study, a cross sectional survey design was used at farm level to collect information from two contrasting agroecological zones through a multistage sampling technique. The AEZ's in terms of counties and sub counties respectively, were selected on the basis of population pressure per square kilometre (high density > 600, medium density 400–599, and low density < 400), rainfall amount and variability as factors that influence climate change and livelihood vulnerability severity impacts. The sampling frame consisted of a list of farmers from target villages provided by the department of agricultural extension, Likuyani and Mavoko sub counties of Kakamega and Machakos counties respectively. Proportionate stratified random sampling was employed with AEZ used as proxy for water availability, use strategies and salinisation risks in the first stage, hence Machakos and Kakamega counties. During the second stage, population density as a proxy for land subdivision (land size), and therefore the extent of land resource marginalization, was used to select villages where the questionnaires and soil sampling were to take place. The third and final stage employed irrigation typology and water source for irrigation. Households for the administration of the questionnaires were then picked through lottery system from a box of cards with numbers generated from a table of random numbers. The semi structured questionnaire was administered between December 2018 and February 2019. The information from household surveys were triangulated through key informant interviews (KI) and focus group discussions (FDGs).

Desk reviews on climate change adaptation policies and environmental governance was also undertaken. Before data collection commenced, the survey questionnaire was tested among 10 respondents to ensure the adequacy of the information obtained and to avoid any ambiguity in the questions. The questionnaire sought information on farmer risk reduction measures concerning soil and water soil testing and associated factors around dissemination of information on salinisation risks. Systematic sampling was employed in the collection of soil and water samples (i.e., on basis of whether ground water (e.g., shallow well, borehole) or surface water (e.g., rivers, roof harvesting) was the main source of irrigation water. Both top soil (0–20 cm) and subsoils (20–40 cm) from irrigated and non-irrigated sections of farmers' fields were collect using a soil auger, packed and analysed through AAS (atomic absorption spectrophotometer) and flame photometer at the Kenya Agricultural and Livestock Research (KARLO), Kabete, an ISO/IEC17025 accredited laboratory. This involved composite sampling where top and subsoil subsamples (four) from each farm and sampling point (zigzag transect) were combined to make up a single composite sample. Composite sampling control for spatial and horizontal variations and improves the accuracy in estimation of population parameters which reduces cost and analytical time [112]. It was assumed that each sample contributes an equal amount to the composite sample and the interaction between the sample units would not significantly affect the eventual composite sample.

3.3. Sample Size Determination

The study employed Fishers formula [113] in the determination of sample size (Equation (1)).

$$n = Z^2 \left[\frac{pq}{d^2} \right] \tag{1}$$

where n = desired sample size, Z = Standard normal deviate at 95% level of confidence = 1.9, P = proportion of target population estimated to have the characteristic under investigation (10% or 0.1) to maximize sample size (precision), q = proportion of target population without the characteristic (1 − p = 90% or 0.9), d= level of precision corresponding to statistical significance level of 0.05 or 5%. Substituting for the values n = Z^2 (p q)/d^2) = $1.96^2(0.1^* 0.9)/(0.05)^2$ = 138.28, hence 139 farmers.

Though the desired sample size from Fishers formula (Equation (1)) is 139 households, we adopted fast and frugal heuristics logic to reduce the sample size to 69. In risk analysis, fast and frugal heuristics logic is normally adopted in cases where ignoring some information does not compromise accuracy of the findings [29]. Given that FGDs and KI interviews carried out a priori revealed farmers across the board in the two counties used similar irrigation practices and tended to have similar dispositions about environmental risks, we adopted the same logic to settle at 50% of the desired sample to maximise precision. The use multistage stratified sampling further justified use of reduced sample size.

3.4. Data Analysis

Statistical analysis was performed using generalised linear logistic weight estimation procedure in IBM[R] SPSS[R] statistics version 26.0 (SPSS Inc., Chicago, IL, USA). A weight estimation procedure computes the coefficients of a linear regression model using weighted least squares (WLS). This ensures that more precise observations (that is, those with less variability) are given greater weight in determining the regression coefficients [114]. WLS thus tests a range of weight transformations that best fit the data. Accordingly, the coefficients selected are those that make the observed results most likely. The weights can be interpreted as a change in the logarithm of the odds ratio E(β), associated with a one-unit change in any predictor. The odds equation is given in Equation (1). A negative E(β)

suggests a decreasing likelihood of falling into the target group as you increase predictor variable, while a positive $E(\beta)$ indicates an increasing likelihood of falling into target group as you increase predictor variable.

$$\Omega = ez/(1 + ez) \tag{2}$$

where Ω is the probability of the event, e is the base of the natural logarithms (2.718), z is the linear combination and calculated as $z = a + \beta_1 x_1 + \beta_2 x_2 + \beta_3 x_3 \ldots + \beta_i x_i$. Whereby, a is a constant (intercept) βs are coefficients (the log odds) estimated from the data and x_i are the values of the predictors.

The log of the odds ratio $E(\beta)$ or logit expression is given as $z = \log (p/(1 - p)$ where P = Probability of falling into target group which is soil/water testing and 1-p = the absence of soil/ water testing on the farm.

Heteroscedasticity which renders estimated ß's inefficient and thus invalid for use in making predictions about dependent variable was tested for through Pearson correlations (Table 4). Though a hypothetical dimension of any social phenomenon can be investigated, the responses may be biased especially where farmers are not familiar with the variable being investigated [115]. Since the occurrence of extreme weather and availability of communication media in extension is common in the two study sites, such biases were controlled for in the study. A hypothetical effect of risk dissemination was thus elicited to visualise if it could bias risk perception in terms of practices that impact salinisation risks. Soil (irrigation water) testing was assumed as the appropriate risk mitigation practice against salinisation. A paired t-test of significance was conducted to evaluate the difference in salinity risks associated with irrigation in the topsoil and subsoil (n = 19) for the two counties. The quantified changes and significance were assumed to provide time scale scenario of salinity and sodium hazard risks.

The sodium adsorption ration (SAR) was calculated according to [116] given in Equation (3)

$$SAR = \frac{Na}{\sqrt{\frac{1}{2}(Ca + Mg)}} \tag{3}$$

where, Na = Sodium in milliequivalents per litre (me/L), Ca = Calcium in milliequivalents per litre (me/L), and Mg = Magnesium in milliequivalents per litre (me/L).

3.5. Student T-Test

The test of significance relating to each regression coefficient of an explanatory variable X_i was made using the t-ratio. The t-statistic (student t test) tests the hypothesis that corresponding independent variables exerts no statistically significant linear influence on dependent variable for the coefficient. It is a ratio of estimated regression coefficient to its standard error (S.E). In general, the null hypothesis is not rejected if the absolute value of t is less than the value of t corresponding to a particular level of significance and it is rejected if the absolute t exceeds this value. A low t-ratio implies that the coefficient is not significant in determining the dependent variable. If, however, the t-ratio exceeds critical value at chosen significance level, then the coefficient is statistically significant. The t-statistic for β_i obtained for a sample is given by Equation (4) while Table 1 provides the independent variables used, their description and their levels

$$t = \frac{\beta i}{\hat{\beta}} \tag{4}$$

where s = standard deviation of the sample and β_i = the estimated value of coefficient.

Table 1. Description of Independent variables and their levels.

Independent variables	Description	Levels
Age AHH	Age of Household head in years	1 = 20–29, 2 = 30–49, 3 = 50–59, 4 = 60–69, 5 = +70
Non-farm income level Household Head (NFIHH)	Level of monthly income from non-farm activities in Kes/Month by the decision maker	1 =< 10,000, 2 = 10,000–19,999, 3 = 20,000–19,999, 4 = 20,000–29,999, 5 = 30,000–39,999, 6 = 40,000–49,999, 7 =>50,000
Farm income household head (FIHH)	Level of gross crop and livestock revenue on the farm (Kes/annum)	1 =< 10,000, 2 = 10,000–19,999, 3 = 20,000–29,999, 4 = 30,000–39,999, 4 = 40,000–49,999, 5 = >50,000
Highest education level of household head (EHH)	Highest level of education attained by the household head	1 = none, 2 = Primary, 3 = Postsecondary but not university, 4 = University and postgraduate
Are you aware of any risks from water source AWR	Knowledge on potential salinisation risks	0 = No, 1 = Yes
Aware of any health risks from water AHR	Main sources of information on health	0 = No, 1 = Yes
Aware of any environmental risks AER	Main source of information on Environmental management	0 = No, 1 = Yes
Believe environmental risks can impart negatively BS	Farmer knowledge on risks associated with salinisation and their effects	0 = No, 1 = Yes
Source of information: environmental SIE	Main source of information on irrigation management	1 = electronic media, 2 = print media, 3 = Private extension, 4 = Public extension, 5 = Radio; Peers = 6, 7 = Journals
Source of information: health SIH	Main source of information on health	1 = electronic media, 2 = print media, 3 = Private extension, 4 = Public extension, 5 = Radio; Peers = 6, 7 = Journals
From whom did you learn about irrigation (L)?	Main sources of information whom the farmer learned irrigation from	1 = electronic media, 2 = print media, 3 = Private extension, 4 = Public extension, 5 = Radio; Peers = 6, 7 = Journals
Types of irrigation (IR)	Types of irrigation technology used on the farm	1 = bucket, 2 = sprinkle, 3 = Drip, 4 = flooding
specific messages on potential risks of different water sources on soil and their control SISST	Whether the farmer received risk messages on salinisation and their control from extension agents	0 = No, 1 = Yes
Source of water for irrigation WS	Main source of water used by the farmer	1 = Spring and Rivers, 2 = Ground (borehole or shallow well, 3 = rain harvesting
Experience with irrigation (TT)	No. of years the farmer has practiced irrigation	1 = 1–4, 2 = 5–9, 3 = 10–15, 4 => 15

Source: Authors conceptualisation, 2019.

4. Results and Discussion

The mean parameters of sampled water used in irrigation for the study area is given in Table 2. There was no significance difference in the parameters between the two counties. The mean hydrogen potential (pH) was 7.2 ± 0.85 with Machakos being 8.37 ± 0.789667 and Kakamega 6.791667 ± 0.263197. Though the mean pH is within the recommended range for most crops, the mean value for Machakos tended towards alkaline with potential to increase salinisation risks. The highest variation in analysed parameters was for chloride levels at 2.4275 ± 14.89418, with Machakos at 7.91 ± 20.98813 accounting for the highest variation. This is expected as the water sources used varied widely from surface to ground water. The source of water is thus critical with ground water (borehole/shallow wells) in Machakos tending to account for extreme chloride values. The mean SAR levels for Machakos were 1.94 ± 8.176467 which increases salinisation risks, while the low levels for Kakamega at 0.14 ± 0.000418 posed low salinisation risks.

Table 2. Mean parameters of water used in irrigation, Machakos and Kakamega counties, Kenya.

Parameter	Overall	Machakos	Kakamega
Ph	7.18625 ± 0.849172	8.37 ± 0.789667	6.791667 ± 0.263197
EC (ms/cm)	0.491875 ± 0.564456	1.605 ± 0.6099	0.120833 ± 0.002627
Na(me/L)	0.246875 ± 0.203956	0.84 ± 0.3936	0.049167 ± 0.000208
K(me/L)	0.02625 ± 0.000452	0.055 ± 0.0000433	0.016667 ± 9.7E-05
Ca(me/L)	0.335 ± 0.8168	1.125 ± 2.9647	0.071667 ± 0.00267
Mg(me/L)	0.42625 ± 0.351265	1.1225 ± 0.798892	0.194167 ± 0.026081
Carbonate(me/L)	0.0275 ± 0.0065	0.11 ± 0.0204	0.000
Bicarbonate(me/L)	0.373125 ± 0.053236	0.7 ± 0.060467	0.264167 ± 0.004299
Chloride(me/L)	2.4275 ± 14.89418	7.91 ± 20.98813	0.6 ± 0.012727
sulphate(me/L)	0.785 ± 0.497907	1.1 ± 1.759267	0.68 ± 0.151055
SAR	0.59 ± 2.2836	1.94 ± 8.176467	0.14 ± 0.000418

Source: Authors statistical analysis of water samples, 2019.

The statistical analysis for changes in sodium in the topsoil (ESP), an indicator of soil salinity hazards is given in Table 3. There is significant difference in salinity hazards in the top soil with irrigation especially for soils in Kakamega county study site. The mean ESP in top and subsoil in Kakamega was 5.65 ± 3.73 and 5.91 ± 0.70 Me% respectively. The mean change in ESP was significant in both sites. The ESP for Kakamega changed by 0.66 ± 0.73 and −0.08 ± 0.40 Me% in top and subsoil respectively. The mean change in ESP for Machakos's study site was 0.033± 0.47and 2.22 ± 28.21 Me% in top and subsoil respectively. The overall change for the two sites with irrigation in the topsoil and subsoil was 0.45 ± 0.70 and 0.69 ± 9.8 Me respectively. The overall negative changes in ESP values for top soil imply displacement or desorption of calcium (ca^{++}), Potassium (k^{+}) and Magnesium ($mg{+}{+}$), the bases that jointly determine cation exchange capacity (CEC), an indicator of soil fertility levels, as more of Na+ is being adsorbed on the soil colloids. The increase in Na^{+} is indicative of soil degradation in terms of dispersion, poor permeability and loss of soil structure risks. The net negative change (decrease) in topsoil is indicative of soil degradation risks while the positive changes (increase) in subsoil soils is attributed to leaching of salts and potentially the degradation of underground water resources over a long planning horizon.

The increase in subsoil Na^{+} levels for both sites could be attributed to leaching of salts under irrigation with high variation in Machakos (2.22 ± 28.21) reflecting the high SAR levels, as well as the high variability of the parameter in water sources (Table 2) utilised for irrigation. The overall mean ESP for both sites was 4.2 with a change of 0.45 in top soil and 4.56 me% in subsoil, a change of 0.69 me%. Overall, irrigation increased ESP in both sites (Table 4), an indicator of soil degradation risks. The two sample F test for variance in Sodium concentration is negative, an indication of increased and high sodicity risks in Kakamega. Though primary salinisation effects were not determined, the increase in sodium concertation with irrigation is indicative of soil quality degradation risks in autonomous adaptation.

Table 3. Paired Two Sample test for Means of ESP (me%) and changes with irrigation (N = 19).

		Treatment	Mean	T	Critical t-Value	Sig
Kakamega (n = 13)	Topsoil	Non-irrigated	5.6 ± 3.73	−7.57	1.8	**
		Change with Irrigation	0.66 ± 0.73			**
	Subsoil	Non-irrigated	5.91 ± 0.70	−24.65	1.8	**
		Change with Irrigation	−0.08 ± 0.40			
Machakos (n = 6)	Topsoil	Non-irrigated	1.4 ± 0.428	−3.69	2.01	**
		Change with Irrigation	0.033 ± 0.47			
	Subsoil	Non-irrigated	1.85 ± 0.84	0.201	2.01	**
		Change with Irrigation	2.22 ± 28.21			
Overall (n = 19)	Topsoil	Non-Irrigated	4.2 ± 6.69	−6.21	1.74	**
		Change with Irrigation	0.45 ± 0.70			
	Subsoil	Non-Irrigated	4.56 ± 4.56	−4.12	1.74	**
		Change with Irrigation	0.69 ± 9.8			

Source: Authors Statistical analysis of Soil samples, ** Significant at 0.05 and 0.001.

Table 4. Two-Sample F-Test for Variance in topsoil sodium concentration (SAR) with irrigation.

	Kakamega	Machakos
Mean	−0.00769	0.133333
Variance	0.001	0.063
Observations	13	6
Df	12	5
F	0.012275	
P(F <= f) one-tail	7.76E-09 **	
F Critical one-tail	0.32197	

Source: Authors analysis of soil and water laboratory statistical analysis, 2019; ** significant at 0.05%.

Table 5 presents Pearsons correlation on a number of factors influencing soil testing in the two study counties. There is a positive correlation between education and income, awareness on risks on water, as well as the positive risk reduction inform of soil/water testing. However, age has a negative correlation on soil (water) testing and salinisation risk reduction. Nonetheless, the more aged believe environmental risks could negatively impart them. Age is also negatively correlated to source of information. Possibly, old farmers tend to rely more on informal sources of information, such as their peers and not the ubiquitous electronic and mass media sources. Age is also negatively correlated with income suggesting that it may constraint adoption of soil testing advisories. In absence of risk communication messages, all the predictors (Table 5) are statistically insignificant in salinisation risk reduction.

Human capital theory [117], identifies innovative ability as closely related to education level, farming experience (proxy for age), and information accumulation. The positive effect observed for education on the adoption of soil testing though not significant is consistent with human capital theory in Agriculture. However, the negative correlation between the number of years spent using of technology (an indirect proxy for age) and perception of harm from environmental risks is consistent with risk normalisation theory [87]. The choice of channels of communication and their effectiveness is thus a critical policy consideration in transformative adaptation and sustainability discourses.

Table 5. Pearson correlations on factors influencing soil testing in Kakamega and Machakos counties, Kenya.

	(Intercept)	Age	NFIHH	FHH	EHH	AWR	WS	AHR	AER	SIH	BS	SIE	L	SISST	IR
(Intercept)	1.000														
Age	−0.416	1.000													
Non-farm income level household head (NFIHH)	−0.320	−0.264	1.000												
Farm income household head (FIHH)	−0.536	0.194	0.377	1.000											
Highest education level - house head (EHH)	0.131	0.476	−0.926	−0.31	1.000										
Are you aware of any risks from water source (AWR)	0.075	−0.781	0.393	−0.195	−0.566	1.000									
Source of water (WS)	−0.568	0.399	−0.210	0.221	0.381	−0.234	1.000								
aware of any health risks from water (AHR)	−0.348	−0.567	0.537	0.100	−0.622	0.611	−0.141	1.000							
aware of any environmental risks (AER)	−0.299	−0.557	0.065	0.054	−0.202	0.640	0.159	0.632	1.000						
Source of information- health	−0.522	−0.269	0.667	0.124	−0.627	0.521	0.162	0.686	0.479	1.000					
believe environmental risk can impart negatively (BS)	−0.048	−0.064	−0.213	−0.219	0.130	0.164	0.336	0.044	0.281	0.076	1.000				
Source of information: environmental (SIE)	−0.473	−0.027	0.668	0.445	−0.647	0.222	0.013	0.427	0.153	0.575	0.302	1.000			
From whom did you get to learn about irrigation (L)?	−0.080	0.399	−0.469	0.044	0.534	−0.379	0.062	−0.424	−0.185	−0.484	−0.406	−0.550	1.000		
specific messages on potential risks of different water sources on soil and their control (SISST)	−0.004	0.643	0.095	0.062	0.127	−0.587	0.042	−0.539	−0.860	−0.262	−0.418	−0.055	0.244	1.000	
Types of irrigation (IR)	−0.115	−0.605	0.057	−0.089	−0.242	0.698	−0.102	0.542	0.878	0.343	0.185	−0.004	−0.071	−0.877	1.000

Source: Authors statistical analysis of field data, 2019.

The communication perspective is critical in risk dissemination and sustainability discourses in climate change adaptation [117–119]. Information improves farmer's human capital, reduces risk and uncertainty in technology adoption process [120]. In this study, the negative correlation between information source and education in risk reduction behaviour is possibly related to biased access of information as the level of education increases. Further, the findings suggest a gap in the current research-extension linkages where access to information sources, such as scientific journals that are more likely to disseminate information on environmental externalities as opposed to the conventional sources, such as the radio are by default biased towards farmers with high levels of education. Since the effect of risk dissemination is negatively correlated with source of information, it suggests that the current sources of information are ineffective and/or do not disseminate information concerning the existing risks. Implicit in this is the need for transformative lenses to enhance the role of media, both electronic and print in risk information dissemination especially as it relates to secondary risks in climate change adaptation.

Table 6 provides the odds ratio E(β), generalised logistic parameter estimates on soil testing as a risk reduction measure and control of irrigation related risks. An odds ratio less than one connotes that the variable decreases the likelihood of adoption, whereas an odds ratio greater than one means that the variable increases the likelihood of adoption. The likelihood of the odds ratio on age, farm income (farm and non-farm), number of years in use of technology, and source of information, education, awareness on health risks, type of, irrigation though not statistically significant had negative odds ratios. In the absence of risk message dissemination, there is a decreased likelihood of soil testing with increase in value of the mentioned variables. From existing literature, risk aversion increases with age hence the negative sign for age in our study is expected. However, education, income, and experience tend to be positively correlated with adoption. This observation suggests that existing technology diffusion and adoption models and human capital theory in agriculture cannot be used effectively to address environmental externalities in adaptation planning.

Table 6. Generalised linear logistic parameter estimates on soil testing without dissemination of risk messages.

Parameter	B	Std. Error	Unstandardized 95% Wald Confidence Interval				Standardized 95% Wald Confidence		
			Lower	Upper	Wald χ^2	Sig.	Exp(β)	Lower	Upper
(Intercept)	−22.572	1.6028	−25.714	−19.431	198.329	0.000	1.574E-10	6.803E-12	3.642E-9
Age	−0.052	0.2414	−0.525	0.421	0.046	0.830	0.950	0.592	1.524
NFIHH	−0.075	0.2073	−0.481	0.332	0.130	0.719	0.928	0.618	1.393
FIHH	−0.110	0.1665	−0.436	0.217	0.433	0.510	0.896	0.647	1.242
EDHH	0.186	0.3147	−0.431	0.803	0.350	0.554	1.205	0.650	2.232
AWR	0.082	0.8013	−1.488	1.653	0.011	0.918	1.086	0.226	5.221
SW	4.855E-5	0.0899	−0.176	0.176	0.000	1.000	1.000	0.838	1.193
AHR	−0.224	1.0522	−2.286	1.838	0.045	0.832	0.799	0.102	6.287
AER	−0.414	0.7711	−1.926	1.097	0.289	0.591	0.661	0.146	2.996
SIH	−0.033	0.0847	−0.199	0.133	0.154	0.695	0.967	0.819	1.142
BS	−0.089	0.4321	−0.936	0.758	0.042	0.837	0.915	0.392	2.134
SIE	−0.003	0.0738	−0.148	0.141	0.002	0.966	0.997	0.863	1.152
L	−0.027	0.0715	−0.167	0.114	0.138	0.711	0.974	0.846	1.120
SSISST	0.068	0.7380	-1.379	1.514	0.008	0.927	1.070	0.252	4.547
IR	0.001	0.1675	−0.327	0.329	0.000	0.995	1.001	0.721	1.390
SI	0.082	0.2351	−0.379	0.543	0.121	0.727	1.085	0.685	1.721
TT	0.004	0.0694	−0.132	0.140	0.003	0.957	1.004	0.876	1.150

Source: Authors statistical analysis of field data, 2019. Likelihood Ratio Chi-Square (χ^2) = 10.858; p = 0.286, df = 9.

The positive effect of risk message dissemination on risk behaviour has been observed by several authors [21,63,86]. The generalised linear logistic parameter estimates (Table 8) explains the effect of risk message dissemination on soil testing. In this study, dissemination of risk messages could have significant impact on likelihood of positive change on risk belief and mitigation action. This is consistent with some findings on rapid onset disasters, such as earthquakes where higher education levels, higher income and greater experience with previous emergencies is significantly associated with higher preparedness [121]. In our study, risk message dissemination has positive significant effect on farmers disposition about salinisation risks with majority of the farmers who would change their behaviour (adopt soil testing as a risk reduction measure) falling in the 30–49 year age category (Table 7).

Table 7. Proportion of change in action for soil testing if risk message were disseminated.

Age Category	No	Yes	Total	% Change
20–29	5	2	7	3.13
30–49	15	18	33	28.13
50–59	8	6	14	9.38
60–69	3	6	9	9.38
70 and above	0	1	1	1.6
Total	31	33	65	51.62

Source: Authors analysis of field data.

Likewise, according to Table 8, dissemination of risk message has significant positive impact on likelihood in change of choice of water sources (WS) for irrigation and type of irrigation (i.e., bucket, sprinkle, surface and drip), all which impact salinity hazards. Additionally, risk message dissemination significantly increases the likelihood of soil testing for every additional level (higher level) of farmer education and the positively correlated non-farm income. However, dissemination of risk messages decreases the likelihood in soil testing when awareness on water and environmental risks are taken into account. This could be due to other factors, notably the extra costs incurred in soil testing as source of risk that decreases profit levels in the short term. The observation is consistent with [69], that gaps between information dissemination and level of implementation could be as a result of subjective limits or considerations for factors that impact profit and/or cost in adoption of risk reduction behaviour. Factors that lower profits or increase expenses are sources of risk (i.e., technical, price, legal, social and human), that adversely impact the economic performance hence farmers' decision making [121–124]. The finding underscores Howden et al. [125], and Koundouri et al. [120], that policy makers in adaptation planning need to increase their attention on the role of risk attitude in technology adoption.

Table 8. Generalised Linear logistic Parameter Estimates on soil testing with dissemination of risk messages.

Parameter	B	Std. Error	Unstandardized 95% Wald Confidence Interval				Standardized 95% Wald Confidence Interval		
			Lower	Upper	Wald χ^2	Sig.	Exp(β)	Lower	Upper
Intercept	−84.523	4.1365	−92.631	−76.416	417.521	0.000	1.959E-37	5.902E-41	6.502E-34
Age	2.782	1.0189	0.785	4.779	7.454	0.006	16.148	2.192	118.961
NFIHH	9.137	0.7023	7.760	1.513	169.256	0.000	9291.669	2345.799	36,804.136
FIHH	−1.196	0.3775	−1.936	−0.457	10.045	0.002	0.302	0.144	0.633
EHH	0.642	0.9184	−1.158	2.442	0.488	0.485	1.899	0.314	11.491
AWR	−9.560	2.4241	−14.311	−4.809	15.553	0.000	7.052E-5	6.094E-7	0.008
WS	0.889	0.1521	0.591	1.187	34.195	0.000	2.434	1.806	3.279
AHR	7.723	2.4725	2.877	12.569	9.755	0.002	2258.738	17.752	287,391.934
AER	−9.136	1.8365	−12.735	−5.537	24.748	0.000	0.000	2.945E-6	0.004
SIH	0.753	0.2005	0.360	1.146	14.085	0.000	2.123	1.433	3.145
BS	7.058	0.7838	5.522	8.594	81.096	0.000	1162.039	250.086	5399.470
SIE	0.228	0.1929	−0.150	0.606	1.400	0.237	1.256	0.861	1.834
L	−0.519	0.1158	−0.746	−0.292	20.089	0.000	0.595	0.474	0.747
SISST	4.927	1.3643	2.253	7.601	13.040	0.000	137.927	9.513	1999.787
IR	2.477	0.4175	1.659	3.295	35.207	0.000	11.908	5.254	26.990
SI	0.353	0.6015	−0.825	1.532	0.345	0.557	1.424	0.438	4.629
TT	−0.618	0.1765	−0.964	−0.272	12.251	0.000	0.539	0.381	0.762

Source: Authors statistical analysis of field data, 2019. Likelihood Ratio Chi-Square (χ^2) = 1.742E^{10}, Df = 7; P = 0.000 ***; significant at 0.001%.

The significant decrease in likelihood of soil testing with risk message dissemination when the number of years the farmer has used a given irrigation technology is taken into account could be attributed to resource fixity in agricultural production (i.e., difficulty in changing irrigation infrastructure to alternative uses) and attendant risks and/or low risk belief about salinisation risks among farmers. The observation is also consistent with existing literature on determinants of cognitive bias, such as, personal experience, knowledge (level of education), extension education, which individually or severally impact cognitive ability and the accuracy of climate information processing [82]. The inherent social and environmental costs in maladaptive projects and their premature decommissioning at a future date may impose high opportunity costs to society at large when adaptation policy and practice ignores the integration of environmental spillover mitigation into planning. The observation highlights the need for system approach and innovative use of communication as a tool for proactive risk reduction and effective adaptation planning.

Managing environmental risks in climate change action inadvertently touches on governance in terms of roles, availing of relevant information, policy and legislative frameworks, risk control guidelines, as well as, coordination mechanism that are responsive to the present and future needs of society [81]. The role of governance on soil testing as a risk management strategy was undertaken through KI, FGDs and desk reviews. The findings revealed key governance gaps, particularly fragmented approaches and coordination among government agencies, low awareness about salinisation risks among farmers and extension agencies, all of which constitute cognitive failure about environmental spillovers in climate change adaptation. Though the object of the climate action planning is to integrate climate risk and vulnerability assessment into all forms of assessment, and for that purpose, to liaise with relevant lead agencies for their technical advice, it tends to focus only on methane emissions and fail to acknowledge the diverse array of environmental spillovers, such as the salinisation risks in irrigation.

In the study area, a lack of coordinated approaches among various agencies was noted. Further, interviews with farmers and analysis of KI interviews revealed that neither the climate change Act nor EMCA identifies salinisation externalities. The cognitive failure was more apparent in extension agencies from both counties. According to KI interviews, the extension agents were more focused on supply and demand needs with irrigation, a routine adjustment and solution to increasingly risky rain fed systems, being recommended to the exclusion of underlying environmental concerns. This seems to be a popular discourse among policy makers, farmers and practionneers in the country.

Some of the projects are funded by the central and county governments against tight timelines, for example emergence drought recovery interventions which tend to be accorded high attention by the political class. We focus on technological dimensions, that is, the agronomic aspects, such as fertilizer types, choice of variety and which are farmer felt needs, but not the environmental spillovers. In any case we have not been notified of any environmental breaches by NEMA agricultural extension officers in the two counties.

The above finding suggest low institutional awareness and fragmented approach, a finding that is consistent with Seidler et al. [6] and Ayers et al. [12], respectively, on determinants of adaptation failure. In addition, an extension officer, Machakos county, had this to say:

"The farmers have not reported any problems with water sources for irrigation except for one borehole in the neighbourhood ... We suspect salinity issues but so far we haven't verified whether the borehole was unsuitable for irrigation or the abandonment was due to other causes"—An agricultural extension officer, Machakos County.

Analysis of water sample from the above-mentioned borehole revealed extremely high salinity and its unsuitability for irrigation. In absence of robust mitigation measures suggested by FAO [33], such as annual soil testing, mixing of rain and borehole water sources, adequate drainage as well as deep tillage, drainage canals, application of manure in large amounts to improve infiltration rate and/or planting crops with good salt tolerance being instituted, there is an increased risk in salinisation and land degradation. Of great concern among surveyed farmers (Table 9) was the widespread ignorance about salinity risks from water sources and their mitigation. The observation is reflective of high level of cognitive failure on soil testing as a risk reduction measure among small scale farmers and government agencies in the two counties. Of the surveyed households, a majority (about 98%) had not undertaken soil testing, with less than 10% of the farmers being aware of salinisation risks. There is a gap in awareness and mitigation. Risk aversion seems to be the explanation for the gap. The farmers had this to say;

Table 9. Farmers undertaking soil testing as a risk reduction measure in Kakamega and Machakos counties, Kenya.

Age	No	Yes	Total	% Testing Soil
20–29	7	0 (0)	7	0 (0)
30–49	33	1 (3)	34	1.54 (4.6)
50–59	14	0 (2)	14	0 (3.1)
60–69	9	0 (1)	9	0 (1.54)
70 and above	1	0 (0)	1	0 (0)
Total	64	1 (6)	65	1.54 (9.24)

Source: Authors statistical analysis of survey data. Figures in brackets indicate those who are aware about risks from water (salinisation risks).

"The frequent droughts have negatively affected our livelihoods yet our ability to respond to it is heavily constrained as we have low incomes. We don't think there are environmental risks other than the problematic pests and diseases that trouble us. If there were environmental risks, we would have heard from some of the extension programmes on radio and the extension officers who rarely visit our farms. In any case we think it could be costly testing the soil and water unless the relevant government agencies provide such services for free"—Farmer FGDs in Kakamega and Machakos counties.

The cognitive failure across individuals and institutions in adaptation planning in the study area reflect the governance gaps about environmental externalities. The pervasiveness of cognition failure, as manifested through low awareness among farmers and government agencies alike, as well as poor coordination among formal agencies especially agricultural extension services, is indicative of ineffective adaptation planning frameworks in the counties and the country at large.

Mu et al. [69], explains the variance between awareness and implementation in terms of profit motives. This may account for the observed negative odds likelihood between risk message dissemination on choice of water source for irrigation. The negative likelihood has profound policy implication and the management of underlying risks, such as the environmental spillovers. Though the risk reduction focused climate change Act has potential to address some of the demand-supply needs and production risks, it fails to recognise the negative environmental spillovers. The cognitive failure is reflected in low institutional attention accorded to slow onset disasters in the NAPAs among lead and regulatory agencies. For example, salinisation risks were not mentioned nor captured as concerns that need monitoring. The cognitive failure is aptly reflected in a lack of mention of salinisation risks and their mitigation in the Environmental Management Plan (EMP) section of environmental impact assessments (EIAs) reports on irrigation undertaken nationally and the study sites.

5. Conclusions

Poor system integration, as well as low attention to spillover systems across scale, especially the low attention to time related integration needs in adaptation planning has potential to exacerbate less recognised slow onset disaster risks, such as salinisation. In absence of a transformative and system approach, failure to identify and internalise the individual and cumulative impacts of the seemingly minor footprints could over time substantially increase land degradation risks and impose costs on the society at large. In this study we explored farmer perception on slow onset disasters and how it constraints transformative adaptation. Specifically, the role of cognition or perception in mobilising peoples' commitment to action over negative environmental externalities, risk belief and mitigation action has been highlighted. The findings suggest that multifaceted biases and failures about the existence and importance of externalities across scale, a critical gap in adaptation planning discourses, is exacerbated through low awareness, fragmented approaches and technological biased lenses among actors in adaptation planning.

Under diverse social-economic contexts education level, farming experience, and information accumulation as human capital components significantly account for adoption of technologies in conventional technology diffusion trajectories. However, from this study, the human capital components do not significantly influence risk reduction behaviour concerning environmental spillovers in absence of risk message information. The failure by diverse actors across scale to recognise the externalities, as well as the low institutional awareness constitute cognitive failure with potential to undermine ecosystems, farmer adaptive capacity and livelihoods in the long run. Transformative adaptation policy framing and information support frameworks have great potential to guide informed decision making and a paradigm shift towards effective adaptation action, learning and mitigation of environmental externalities. This is particularly relevant for slow onset disasters, such as salinisation related land degradation risks, where lack and /or poor knowledge of the consequences of the effect resonates with the narrative of wicked environmental problems and adaptation failure. Electronic and print media could compliment conventional extension strategies in risk information dissemination, especially as relates to the mitigation of secondary risks in climate change adaptation.

Author Contributions: V.T.E. conceived the basic idea of the study. V.T.E. and J.O.O. designed the structure of the study. V.T.E. wrote while J.O.O. revised the manuscript. Both authors approved the final version of manuscript. All authors have read and agreed to the published version of the manuscript.

Funding: This research received no external funding.

Acknowledgments: We acknowledge farmers in Likuyani and Mavoko sub counties and respective extension staff, Ministry of Agriculture whose support enabled us to collect the relevant data, as well as the comments of three anonymous reviewers who greatly contributed to the improvement of the manuscript.

Conflicts of Interest: The authors declare no conflict of interest. This research did not receive any specific grant from funding agencies in the public, commercial, or not-for-profit sectors.

References

1. Pelling, M.; O'Brien, K.; Matyas, D. Adaptation and transformation. *Clim. Chang.* **2015**, *133*, 113–127. [CrossRef]
2. Pahl-Wostl, C. A conceptual framework for analysing adaptive capacity and multi-level learning processes in resource governance regimes. *Glob. Environ. Chang.* **2009**, *19*, 354–365. [CrossRef]
3. Rijke, J.; Brown, R.; Zevenbergen, C.; Ashley, R.; Farrelly, M.; Morison, P.; Herk, S. Van Fit-for-purpose governance: A framework to make adaptive governance operational. *Environ. Sci. Policy* **2012**, *22*, 73–84. [CrossRef]
4. United Nations. *Transforming the World: The 2030 Agenda for Sustainable Development*; United Nations: New York, NY, USA, 2015; A/RES/70/1.
5. United Nations International Strategy for Disaster Reduction. *Sendai Framework for Disaster Risk Reduction 2015–2030*; UN: Sendai, Japan, 2015.
6. Seidler, R.; Dietrich, K.; Schweizer, S.; Bawa, K.S.; Chopde, S.; Zaman, F.; Sharma, A.; Bhattacharya, S.; Devkota, L.P.; Khaling, S.; et al. Progress on integrating climate change adaptation and disaster risk reduction for sustainable development pathways in South Asia: Evidence from six research projects. *Int. J. Disaster Risk Reduct.* **2018**, *31*, 92–101. [CrossRef]
7. Liu, J.; Mooney, H.; Hull, V.; Davis, S.; Gaskell, J.; Hertel, T.; Lubchenco, J.; Seto, K.; Gleick, P.; Kremen, C.; et al. Systems integration. *Sustainability* **2015**, *347*, 963–971.
8. Reed, M.S.; Podesta, G.; Fazey, I.; Geeson, N.; Hessel, R.; Hubacek, K.; Letson, D.; Nainggolan, D.; Prell, C.; Rickenbach, M.G.; et al. Combining analytical frameworks to assess livelihood vulnerability to climate change and analyse adaptation options. *Ecol. Econ.* **2013**, *94*, 66–77. [CrossRef]
9. Park, S.E.; Marshall, N.A.; Jakku, E.; Dowd, A.M.; Howden, S.M.; Mendham, E.; Fleming, A. Informing adaptation responses to climate change through theories of transformation. *Glob. Environ. Chang.* **2012**, *22*, 115–126. [CrossRef]

10. Nagoda, S. New discourses but same old development approaches? Climate change adaptation policies, chronic food insecurity and development interventions in northwestern Nepal. *Glob. Environ. Chang.* **2015**, *35*, 570–579. [CrossRef]

11. Berkes, F.; Ross, H. Community Resilience: Toward an Integrated Approach Community Resilience: Toward an Integrated Approach. *Soc. Nat. Resour.* **2013**, *26*, 5–20. [CrossRef]

12. Ayers, J.; Huq, S.; Wright, H.; Faisal, A.M.; Tanveer, S. Mainstreaming climate change adaptation into development in Bangladesh. *Clim. Dev.* **2014**, *6*, 293–305. [CrossRef]

13. Huq, N.; Renaud, F.; Sebesvari, Z. *Ecosytem Based Adaptation (EbA) to Climate Change—Integrating Actions to Sustainable Adaptation*; Institute for Environmental and Human Security, United Nations University: Bonn, Germany, 2013.

14. Wise, R.M.; Fazey, I.; Smith, M.S.; Park, S.E.; Eakin, H.C.; Van Garderen, E.R.M.A.; Campbell, B. Reconceptualising adaptation to climate change as part of pathways of change and response. *Glob. Environ. Chang.* **2014**, *28*, 325–336. [CrossRef]

15. Stafford Smith, M.; Harrocks, L.; Harvey, A.; Hamilton, C. Rethinking adaptation for a 4 °C world. *Philos. Trans. R. Soc. A* **2011**, *369*, 196–216. [CrossRef] [PubMed]

16. Nelson, D.R.; Adger, W.N.; Brown, K. Adaptation to Environmental Change: Contributions of a Resilience Framework. *Annu. Rev. Environ. Resour.* **2007**, *32*, 395–419. [CrossRef]

17. Dupuis, J.; Knoepfel, P. The Adaptation Policy Paradox: The Implementation Deficit of Policies. *Ecol. Soc.* **2013**, *18*, 31–47. [CrossRef]

18. Stirling, A. Opening up or closing down? Analysis, participation and power in the social appraisal of technology. In *Science and Citizen: Globalization and the Challenge of Engagement*; Leach, M., Scoones, L., Wynne, B., Eds.; ZeD Books: London, UK, 2005; pp. 218–231. ISBN 1842775502.

19. Brown, K. Sustainable adaptation: An oxymoron? *Clim. Dev.* **2011**, *5529*, 20–31. [CrossRef]

20. Schipper, L.; Pelling, M. Disaster risk, climate change and international development: Scope for, and challenges to, integration. *Disasters* **2006**, *30*, 19–38. [CrossRef]

21. Adger, W.N.; Dessai, S.; Goulden, M.; Hulme, M.; Lorenzoni, I.; Nelson, D.R.; Otto, L.; Johanna, N.; Anita, W. Are there social limits to adaptation to climate change? *Clim. Chang.* **2009**, *93*, 335–354. [CrossRef]

22. Gostin, L.; Lucey, D.; Phelan, A. The Ebola Epidemic A Global Health Emergency. *JAMA* **2014**, *312*, 1095–1096. [CrossRef]

23. Pelling, M.; Manuel-Navarrete, D. From Resilience to Transformation: The Adaptive Cycle in Two Mexican Urban Centers. *Ecol. Soc.* **2011**, *16*, 11–22. [CrossRef]

24. Ostrom, E. A diagnostic approach for going beyond panaceas. *Proc. Natl. Acad. Sci. USA* **2007**, *104*, 15181–15187. [CrossRef]

25. Paavola, J.; Gouldson, A.; Kluvánková-Oravská, T. Interplay of Actors, Scales, Frameworks and Regimes in the Governance of Biodiversity. *Environ. Policy Gov.* **2009**, *158*, 148–158. [CrossRef]

26. Karpouzoglou, T.; Dewulf, A.; Clark, J. Advancing adaptive governance of social-ecological systems through theoretical multiplicity. *Environ. Sci. Policy* **2016**, *57*, 1–9. [CrossRef]

27. Liu, J.; Yang, W. Integrated assessments of payments for ecosystem services programs. *Proc. Natl. Acad. Sci. USA* **2013**, *110*, 16297–16298. [CrossRef] [PubMed]

28. Suckall, N.; Tompkins, E.; Stringer, L. Identifying trade-offs between adaptation, mitigation and development in community responses to climate and socio-economic stresses: Evidence from Zanzibar, Tanzania. *Appl. Geogr.* **2014**, *46*, 111–121. [CrossRef]

29. Hafenbrädl, S.; Haven, N.; States, U.; Waeger, D.; Marewski, J.N. Applied Decision Making With Fast-and-Frugal Heuristics. *J. Appl. Res. Mem. Cogn.* **2016**, *5*, 215–231. [CrossRef]

30. Reed, M.S.; Fraser, E.D.G.; Dougill, A.J. An adaptive learning process for developing and applying sustainability indicators with local communities. *Ecol. Econ.* **2006**, *59*, 406–418. [CrossRef]

31. Reid, P.; Coleen, V. Living and responding to multiple stressors in South Africa—Glimpses from KwaZulu-Natal. *Glob. Environ. Chang.* **2006**, *16*, 195–206. [CrossRef]

32. Metternicht, G.I.; Zinck, J.A. Remote sensing of soil salinity: Potentials and constraints. *Remote Sens. Environ.* **2003**, *85*, 1–20. [CrossRef]

33. FAO. *Water Quality for Agriculture*; Food and Agricultue Organisation of the United Nations: Rome, Italy, 1985.

34. Jung, T.; Srinivasan, A.; Tamura, K.; Sudo, T.; Watanabe, R.; Shimada, K.; Kimura, H. *Asian Perspectives on Climate Regime Beyond 2012: Concerns, Interests and Priorities*; Istitute for Global Environmental Strategies: Hayama, Japan, 2005.
35. Haddad, B.M. Ranking the adaptive capacity of nations to climate change when socio-political goals are explicit. *Glob. Environ. Chang.* **2005**, *15*, 165–176. [CrossRef]
36. Füssel, H.M. Adaptation planning for climate change: Concepts, assessment approaches, and key lessons. *Sustain. Sci.* **2007**, *2*, 265–275. [CrossRef]
37. Adger, W.N.; Arnell, N.W.; Tompkins, E.L. Successful adaptation to climate change across scales. *Glob. Environ. Chang.* **2005**, *15*, 77–86. [CrossRef]
38. Barnett, J.; O'Neill, S. Maladaptation. *Glob. Environ. Chang.* **2010**, *20*, 211–213. [CrossRef]
39. Mustafa, D. The Production of an Urban Hazardscape in Pakistan: Modernity, Vulnerability, and the Range of Choice. *Ann. Assoc. Am. Geogr.* **2005**, *95*, 566–586. [CrossRef]
40. Bizikova, L.; Robinson, J.; Cohen, S. Linking climate change and sustainable development at the local level. *Clim. Policy* **2007**, *7*, 271–277. [CrossRef]
41. Butler, J.R.A.; Suadnya, W.; Puspadi, K.; Sutaryono, Y.; Wise, R.M.; Skewes, T.D.; Kirono, D.; Bohensky, E.L.; Handayani, T.; Habibi, P.; et al. Framing the application of adaptation pathways for rural livelihoods and global change in eastern Indonesian islands. *Glob. Environ. Chang.* **2014**, *28*, 368–382. [CrossRef]
42. Adger, N.; Aggarwal, P.; Agrawala, S.; Alcamo, J.; Allali, A.; Cruz, R.; Alcaraz, E.D.A.; Easterling, W.; Field, C.; Fischlin, A.; et al. *Climate Change 2007: Impacts, Adaptation and Vulnerability Working Group II Contribution to the Intergovernmental Panel on Climate Change Fourth Assessment Report*; IPCC: Geneva, Swirtzeland, 2007.
43. Alam, G.M.M.; Alam, K.; Mushtaq, S. Climate change perceptions and local adaptation strategies of hazard-prone rural households in Bangladesh. *Clim. Risk Manag.* **2017**, *17*, 52–63. [CrossRef]
44. Elum, Z.A.; Modise, D.M.; Marr, A. Climate Risk Management Farmer's perception of climate change and responsive strategies in three selected provinces of South Africa. *Clim. Risk Manag.* **2017**, *16*, 246–257. [CrossRef]
45. Niles, M.T.; Lubell, M.; Haden, V.R. Perceptions and responses to climate policy risks among California farmers. *Glob. Environ. Chang.* **2013**, *23*, 1752–1760. [CrossRef]
46. Gordon, A.J.; Morton, L.W.; Hobbs, J. Understanding Farmer Perspectives on Climate Change Adaptation and Mitigation: The Roles of Trust in Sources of Climate Information, Climate Change Beliefs, and Perceived Risk. *Environ. Behav.* **2015**, *47*, 205–234.
47. Lang, T.; Rayner, G. Ecological Public Health. The 21st century's big idea. *BMJ* **2012**, *345*, 17–20. [CrossRef]
48. David, E.; Elise, H. Climate Change: Perceptions and Discourses of Risk Climate Change: Perceptions and Discourses of Risk. *J. Risk Res.* **2007**, *10*, 623–641.
49. UNEP. Avoiding Future Famines: Strengthening the Ecological Foundation of Food Security through Sustainable Food Systems. In *A UNEP Synthesis Report*; UNEP: Nairobi, Kenya, 2012.
50. Fisher, E.; Jones, J.; von Schomberg, R. (Eds.) Implementing the Precautionary Principle: Perspectives and Prospects. In *Implementing the Precautionary Principle*; Edward Elgar: Cheltenham, UK; Northampt, MA, USA, 2006; Volume 136, pp. 1–11.
51. Schindler, B.D.E.; Hilborn, R. Prediction, precaution, and policy under global change. *Science* **2015**, *347*, 953–954. [CrossRef] [PubMed]
52. Shivakoti, G.; Ullah, R.; Pradhan, U. Challenges of Sustainable Natural Resources Management in Dynamic Asia. In *Redefining Diversity and Dynamics of Natural Resources Management in Asia*; Elsevier Inc.: Amsterdam, The Netherlands, 2016; Volume 1, pp. 3–12.
53. Thapa, B.; Scott, C.; Wester, P.; Varady, R. Towards characterizing the adaptive capacity of farmer-managed irrigation systems: Learnings from Nepal. *Environ. Sustain.* **2016**, *21*, 37–44. [CrossRef]
54. Moench, M. Development in Practice Experiences applying the climate resilience framework: Linking theory with practice. *Dev. Pract.* **2014**, *24*, 447–464. [CrossRef]
55. Schipper, E.L.F.; Thomalla, F.; Vulturius, G.; Davis, M.; Johnson, K. Linking disaster risk reduction, climate change and development. *Int. J. Disaster Reslience Built Environ.* **2016**, *7*, 216–228. [CrossRef]
56. Mukheibir, P.; Ziervogel, G. Developing a Municipal Adaptation Plan (MAP) for climate change: The city of Cape Town. *Environ. Urban.* **2007**, *19*, 143–158. [CrossRef]

57. Meek, C.L.; Lauren, A.; Varjopuro, R.; Dowsley, M.; Dale, A.T. Adaptive governance and the human dimensions of marine mammal management: Implications for policy in a changing North. *Mar. Policy* **2011**, *35*, 466–476. [CrossRef]

58. Termeer, C.; Dewulf, A.; Breeman, G. Governance of Wicked Climate Adaptation Problems. In *Climate Change Governance, Climate Change Management*; Knieling, J., Walter, L.F., Eds.; Springer: Berlin/Heidelberg, Germany, 2013; pp. 27–39.

59. Baer, P. Adaptation: Who pays Whom? In *Fairness in Adaptation to Climate Change*; Adger, W.N., Pavoola, J., Huq, S., Mace, M.J., Eds.; MIT Press: Cambridge, MA, USA, 2006; pp. 133–153. ISBN 9780262012270336.

60. Ding, Y.; Peng, J. Impacts of Urbanization of Mountainous Areas on Resources and Environment: Based on Ecological Footprint Model. *Sustainability* **2018**, *10*, 765. [CrossRef]

61. Loboguerrero, A.M.; Campbell, B.M.; Cooper, P.J.M.; Hansen, J.W.; Rosenstock, T.; Wollenberg, E. Food and Earth Systems: Priorities for Climate Change Adaptation and Mitigation for Agriculture and Food Systems. *Sustainability* **2019**, *11*, 1372. [CrossRef]

62. Pahl-Wostl, C. Transitions towards adaptive management of water facing climate and global change. *Water Resour. Manag.* **2007**, *21*, 49–62. [CrossRef]

63. FAO. *The Water-Energy-Food Nexus*; FAO: Rome, Italy, 2014.

64. Paton, D.; McClure, J. *Preparing for Disaster: Building Household and Community Capacity*; Charles C Thomas Publisher Ltd.: Springfield, IL, USA, 2013.

65. Spence, A.; Pidgeon, N. Framing and communicating climate change: The effects of distance and outcome frame manipulations. *Glob. Environ. Chang.* **2010**, *20*, 656–667. [CrossRef]

66. Volenzo, T.E.; Odiyo, J.O. Ecological public health and participatory planning and assessment dilemmas: The case of water resources management. *Int. J. Environ. Res. Public Health* **2018**, *15*, 1635. [CrossRef] [PubMed]

67. Ayers, J.M.; Huq, S. The Value of Linking Mitigation and Adaptation: A Case Study of Bangladesh. *Environ. Manag.* **2009**, *43*, 753–764. [CrossRef] [PubMed]

68. Pelling, M.; High, C. Understanding adaptation: What can social capital offer assessments of adaptive capacity? *Glob. Environ. Chang.* **2005**, *15*, 308–319. [CrossRef]

69. Mu, D.; Kaplan, T.R.; Dankers, R. Decision making with risk-based weather warnings. *Int. J. Disaster Risk Reduct.* **2018**, *30*, 59–73. [CrossRef]

70. Niles, M.T.; Lubell, M.; Brown, M. Agriculture, Ecosystems and Environment. How limiting factors drive agricultural adaptation to climate change. *Agric. Ecosyst. Environ.* **2015**, *200*, 178–185. [CrossRef]

71. Jackson, L.; Van Noordwijk, M.; Bengtsson, J.; Foster, W.; Lipper, L.; Pulleman, M.; Said, M.; Snaddon, J.; Vodouhe, R. Biodiversity and agricultural sustainagility: From assessment to adaptive management. *Curr. Opin. Environ. Sustain.* **2010**, *2*, 80–87. [CrossRef]

72. Volenzo, T.E.; Odiyo, J.; Obiri, J. Greenhouse gas emissions as sustainability indicators in agricultural sectors ' adaptation to climate change: Policy implications. *J. Disaster Risk Stud.* **2019**, *11*, 9. [CrossRef]

73. Walthall, C.L.; Hatfield, J.; Backlund, P.; Lengnick, L.; Marshall, E.; Walsh, M.; Adkins, S.; Aillery, M.; Ainsworth, E.A.; Ammann, C.; et al. *Climate Change and Agriculture in the United States: Effects and Adaptation*; USDA, Agricultural Research Service and Climate Change Program Office: Washington, DC, USA, 2012.

74. MEA. *Ecosystems and Human Well-Being: A Framework for Assessment*; Island Press: Washington, DC, USA; Covelo, CA, USA; London, UK, 2003.

75. Parish, E.S.; Herzberger, A.J.; Phifer, C.C.; Dale, V.H. Transatlantic wood pellet trade demonstrates telecoupled benefits. *Ecol. Soc.* **2018**, *23*. [CrossRef]

76. Liu, J.; Hull, V.; Batistella, M.; Defries, R.; Dietz, T.; Fu, F.; Hertel, T.W.; Cesar, R.; Lambin, E.; Li, S.; et al. Framing Sustainability in a Telecoupled World. *Ecol. Soc.* **2013**, *18*, 26–45. [CrossRef]

77. Adger, W.N. Social Capital, Collective Action, and Adaptation to Climate Change. *Econ. Geogr.* **2003**, *79*, 387–404. [CrossRef]

78. Holland, T.; Smit, B. Climate Change and the Wine Industry: Current Research Themes and New Directions. *J. Wine Res.* **2010**, *21*, 125–136. [CrossRef]

79. Rogers, M.B.; Amlôt, R.; Rubin, G.J.; Wessely, S.; Krieger, K. Mediating the social and psychological impacts of terrorist attacks: The role of risk perception and risk communication. *Int. Rev. Psychiatry* **2007**, *19*, 279–288. [CrossRef] [PubMed]

80. Easterling, W.; Aggarwal, P.; Batima, P.; Brander, K.; Erda, L.; Howden, M.; Kirilenko, A.; Morton, J.; Soussana, J.F.; Schmidhuber, J.; et al. Food, fibre and forest products. In *Climate Change 2007: Impacts, Adaptation and Vulnerability. Contribution of Working Group II to the Fourth Assessment Report of the Intergovernmental Panel on Climate Change*; Parry, M.L., Canziani, O.F., Palutikof, J.P., van der Linden, P.J., Hanson, C.E., Eds.; Cambridge University Press: Cambridge, UK, 2007; pp. 273–313.

81. Volenzo, T.E.; Odiyo, J.O. Linking risk communication and sustainable climate change action: A conceptual framework. *J. Disaster Risk Stud.* **2019**, *11*, 1–11. [CrossRef]

82. Hitayezu, P.; Wale, E.; Ortmann, G. Climate Risk Management Assessing farmers' perceptions about climate change: A double-hurdle approach. *Clim. Risk Manag.* **2017**, *17*, 123–138. [CrossRef]

83. IPCC. *Managing the Risks of Extreme Events and Disasters to Advance Climate Change Adaptation. A Special Report of Working Groups I and II of the Intergovernmental Panel on Climate Change*; Cambridge University Press: Cambridge, UK, 2012.

84. Matyas, D.; Pelling, M. Positioning resilience for 2015: The role of resistance, incremental adjustment and transformation in disaster risk management policy. *Disasters* **2015**, *39*, s1–s18. [CrossRef]

85. Pidgeon, N.; Rogers-Hayden, T. Opening up nanotechnology dialogue with the publics: Risk communication or "upstream engagement"? *Health Risk Soc.* **2007**, *9*, 191–210. [CrossRef]

86. Jihan, N.; Mcshane, P. Landscape and Urban Planning Ecosystem services management: An evaluation of green adaptations for urban development in Dhaka, Bangladesh. *Landsc. Urban Plan.* **2018**, *173*, 23–32.

87. Becker, J.S.; Paton, D.; Johnston, D.M.; Ronan, K.R.; McClure, J. The role of prior experience in informing and motivating earthquake preparedness. *Int. J. Disaster Risk Reduct.* **2017**, *22*, 179–193. [CrossRef]

88. Dietz, T.; Dan, A.; Shwom, R. Support for Climate Change Policy: Social Psychological and Social Structural Influences. *Rural Sociol.* **2007**, *72*, 185–214. [CrossRef]

89. Aldunce, P.; Beilin, R.; Howden, M.; Handmer, J. Resilience for disaster risk management in a changing climate: Practitioners' frames and practices. *Glob. Environ. Chang.* **2015**, *30*, 1–11. [CrossRef]

90. Ravikumar, P.; Somashekar, R.K. Assessment and Modelling of Groundwater Quality Data and Evaluation of Their Corrosiveness and Scaling Potential Using Environmetric Methods in Bangalore South Taluk, Karnataka state, India. *Water Resour.* **2012**, *39*, 446–473. [CrossRef]

91. Asfaw, E.; Suryabhagavan, K.V.; Argaw, M. Soil salinity modeling and mapping using remote sensing and GIS: The case of Wonji sugar cane irrigation farm, Ethiopia. *J. Saudi Soc. Agric. Sci.* **2018**, *17*, 250–258. [CrossRef]

92. Elhag, M. Evaluation of Different Soil Salinity Mapping Using Remote Sensing Techniques in Arid Ecosystems, Saudi Arabia. *J. Sens.* **2016**, *2016*, 7596175. [CrossRef]

93. van Rensburg, L.; de Clercq, W.; Barnard, J.; Preez Du, C. Salinity guidelines for irrigation: Case studies from Water Research Commission projects along the Lower Vaal, Riet, Berg and Breede Rivers. *Water SA* **2011**, *37*, 739–750. [CrossRef]

94. Zewdu, S.; Suryabhagavan, K.V.; Balakrishnan, M. Geo-spatial approach for soil salinity mapping in Sego Irrigation Farm, South Ethiopia. *J. Saudi Soc. Agric. Sci.* **2017**, *16*, 16–24. [CrossRef]

95. Rietz, D.N.; Haynes, R.J. Effects of irrigation-induced salinity and sodicity on soil microbial activity. *Soil Biol. Biochem.* **2003**, *35*, 845–854. [CrossRef]

96. Wei, Q.; Xu, J.; Liao, L.; Li, Y.; Wang, H.; Rahim, S.F. Water Salinity Should Be Reduced for Irrigation to Minimize Its Risk of Increased Soil N_2O Emissions. *Int. J. Environ. Res. Public Health* **2018**, *15*, 2114. [CrossRef]

97. Gorji, T.; Sertel, E.; Tanik, A. Monitoring soil salinity via remote sensing technology under data scarce conditions: A case study from Turkey. *Ecol. Indic.* **2017**, *74*, 384–391. [CrossRef]

98. Odiyo, J.O.; Makungo, R.; Muhlarhi, T.G. The impacts of geochemistry and agricultural activities on groundwater quality in the soutpansberg fractured aquifers. *WIT Trans. Ecol. Environ.* **2014**, *182*, 121–132.

99. Turner, B.L.; Kasperson, R.E.; Matsone, P.A.; McCarthy, J.J.; Corell, R.W.; Christensene, L.; Eckley, N.; Kasperson, J.X.; Amy, L.; Marybeth, A.L.; et al. A framework for vulnerability analysis in sustainability science. *Proc. Natl. Acad. Sci. USA* **2003**, *100*, 8074–8079. [CrossRef]

100. Bhattacharyya, R.; Ghosh, B.; Mishra, P.; Mandal, B.; Rao, C.S.; Sarkar, D.; Das, K.; Anil, K.S.; Lalitha, M.; Hati, K.; et al. Soil Degradation in India: Challenges and Potential Solutions. *Sustainability* **2015**, *7*, 3528–3570. [CrossRef]

101. Pelling, M.; Özerdem, A.; Barakat, S. The macro-economic impact of disasters. *Prog. Dev. Stud.* **2002**, *2*, 283–305. [CrossRef]
102. Zamani, G.H.; Gorgievski-Duijvesteijn, M.; Zarafshani, K. Coping with Drought: Towards a Multilevel Understanding Based on Conservation of Resources Theory. *Hum. Ecol.* **2006**, *34*, 677–692. [CrossRef]
103. G.O.K. *Agricultural Sector Transformation and Gowth Strategy 2019–2029*; Government printer: Nairobi, Kenya, 2019.
104. G.O.K. *Climate Change Action Plan 2018–2022*; Ministry of Environment and Forestry: Nairobi, Kenya, 2018.
105. Republic of Kenya. *Environmental Management and Co-Ordination (Amendment) Act No. 8 of 1999 (Revised, 2012)*; Kenya Law press: Nairobi, Kenya, 2015; p. 81.
106. Republic of Kenya. *Climate Change Act No. 11 of 2016*; Government printer: Nairobi, Kenya, 2016; pp. 1–15.
107. KNBS. *Counting Our People for Implementation of Vision 2030*; Government Printer: Nairobi, Kenya, 2009.
108. Jaetzold, R.; Schmidt, H.; Hornetz, B.; Shisanya, C. *Farm Management Handbook of Kenya Vol 11—Natural Condtions and Farm Management Information*, 2nd ed.; Part II/A1(western Kenya) and Part II/CI(Eastern Province); Ministry of Agriculture, Kenya/GTZ: Nairobi, Kenya, 2011; Volume II.
109. Republic of Kenya. *Kakamega County Integrated Development Plan 2018–2022*; County Government of Kakamega: Kakamega, Kenya, 2019.
110. FAO. *Agro-Ecological Zoning Guidelines: FAO Soils Bulletin No. 73*; Land and Water Development Division, FAO: Rome, Italy, 1996.
111. Republic of Kenya. *Machakos County Integrated Development Plan, 2015*; County Government of Machakos: Machakos, Kenya, 2015.
112. Chen, T.; Liu, X.; Li, X.; Zhao, K.; Zhang, J.; Xu, J.; Shi, J.; Dahlgren, A. Heavy metal sources identification and sampling uncertainty analysis in a field-scale vegetable soil of Hangzhou, China. *Environ. Pollut.* **2009**, *157*, 1003–1010. [CrossRef]
113. Fisher, A.; Laing, J.; Stoeckel, J. *A Handbook for Family Planning Operations Research Designs*; The Population Council: New York, NY, USA, 1983.
114. Hosmer, D.W.; Lemeshow, S. *Applied Logistic Regression*, 1st ed.; John Wiley and Sons: New York, NY, USA, 2000.
115. Helgeson, J.F.; Dietz, S.; Hochrainer-stigler, S. Vulnerability to Weather Disasters: The Choice of Coping Strategies in Rural Uganda. *Ecol. Soc.* **2013**, *18*, 2–15. [CrossRef]
116. Richards, L.A. *Diagnosis and Improvement of Saline and Alkali Soils*; No. Agricultural Handbook No.60; US Department of Agriculture: Washington, DC, USA, 1954.
117. Huffman, W.E. Human Capital: Education and Agriculture. In *Handbook of Agricultural Economics*; Rausser, G.C., Gardner, B.L., Eds.; Elsevier: New York, NY, USA, 2001; pp. 333–381.
118. Mojtahedi, M.; Oo, B.L. Critical attributes for proactive engagement of stakeholders in disaster risk management. *Int. J. Disaster Risk Reduct.* **2017**, *21*, 35–43. [CrossRef]
119. Evans, H.; Dyll, L.; Teer-tomaselli, R. Communicating Climate Change: Theories and Perspectives. In *Handbook of Climate Change Communication: Climate Change Management*; Leal Filho, W., Manolas, E., Azul, A.M., Azeiteiro, U.M., McGhie, H., Eds.; Springler AG: Berlin, Germany, 2018; Volume 1, pp. 107–122.
120. Koundouri, P.; Nauges, C.; Tzouvelekas, V. Technology Adoption under Production Uncertainty: Theory and Application to Irrigation Technology. *Am. J. Agric. Econ.* **2019**, *88*, 657–670. [CrossRef]
121. Shapira, S.; Aharonson-daniel, L.; Bar-dayan, Y. Anticipated behavioral response patterns to an earthquake: The role of personal and household characteristics, risk perception, previous experience and preparedness. *Int. J. Disaster Risk Reduct.* **2018**, *31*, 1–8. [CrossRef]
122. Bailey, B.K.W. *The Fundamentals of Forward Contracting, Hedging, and Options for Dairy Producers in the Northeast*; Penn State University: State College, PA, USA, 2001.
123. Maredia, M.K.; Minde, I.J. Technology, Profitability and agricultural Transformation: Concepts, Evidence and Policy implications. In *Perspectives on Agricultural Transformation. A View from Africa*; Jayne, T.S., Minde, P., Argwings-Kodhek, G.G., Eds.; Nova Science Publishers Inc.: New York, NY, USA, 2002; pp. 83–116.

124. Ullah, R.; Shivakoti, G.P.; Zulfiqar, F.; Kamran, M.A. Farm risks and uncertainties: Sources, impacts and management. *Outlook Agric.* **2016**, *45*, 199–205. [CrossRef]

125. Howden, S.; Jean-Fancois, S.; Tubiello, F.; Chhetri, N.; Dunlop, M.; Meinke, H. Adapting agriculture to climate change. *Proc. Natl. Acad. Sci. USA* **2007**, *104*, 19691–19696. [CrossRef]

Article

Economic and Ecological Impacts of Increased Drought Frequency in the Edwards Aquifer

Jinxiu Ding [1] and Bruce A. McCarl [2,*]

[1] Department of Public Finance, School of Economics, Xiamen University, Xiamen 361005, China; jinxiuding@xmu.edu.cn

[2] Department of Agricultural Economics, Texas A&M University, College Station, TX 77840, USA

* Correspondence: mccarl@tamu.edu

Received: 13 November 2019; Accepted: 18 December 2019; Published: 20 December 2019

Abstract: This paper examines how increased drought frequency impacts water management in arid region, namely the Edwards Aquifer (EA) region of Texas. Specifically, we examine effects on the municipal, industrial, and agricultural water use; land allocation; endangered species supporting springflows and welfare. We find that increases in drought frequency causes agriculture to reduce irrigation moving land into grassland for livestock with a net income loss. This also increases water transfer from irrigation uses to municipal and industrial uses. Additionally, we find that regional springflows and well elevation will decline under more frequent drought condition, which implicates the importance of pumping limits and/or minimum springflow limits. Such developments have ecological implications and the springflows support endangered species and a switch from irrigated land use to grasslands would affect the regional ecological mix.

Keywords: drought frequency; water use; land conversion; livestock production; ecological implications

1. Introduction

The Edwards Aquifer (EA) provides high-quality water to more than 2 million people in the Texas counties of Kinney, Uvalde, Medina, Bexar, Comal, and Hays and supplies a considerable proportion of the base flow to the Guadalupe River. The EA water supports pumping use by irrigated cropping, households, businesses, industries, and users of spring-fed rivers. Aquifer fed springs provide important habitat for endangered species [1]. The EA water supply relies on precipitation-based recharge, which is highly influenced by weather and adversely affected by drought.

Climate change may alter drought frequency and affect water use in the EA region. The Special Report on Managing the Risks of Extreme Events and Disasters to Advance Climate Change Adaptation (SREX) report of the Intergovernmental Panel on Climate Change (IPCC) shows that changing climate can result in alterations in the frequency, duration, and intensity of extreme weather events [2]. More frequent and severe droughts are expected across many regions in the 21st century [3–6]. Changing climate is projected to reduce groundwater resources in most dry subtropical regions, which may intensify water use competition among sectors [7]. Texas has been projected to experience more frequent future drought [8,9]. Increases in the drought frequency or average temperature along with decreases in rainfall all increase water demand but lower water availability.

Recent climatic trends may increase water and drought concerns. For example, the 2000's was the warmest decade on record [10], and 2011 was the strongest La Niña year on record [11]. Additionally, La Niña years are associated with low EA region recharge and drought [12]. Texas and the EA region have been facing drought conditions for the last few years, and 2011 was the most severe, with 88% of the entire state under exceptional drought conditions in September/October [13].

This study examines the implications of increasing drought frequency on the EA region in terms of water availability, water use, agricultural production, land allocation, springflow, welfare, and ecology.

2. Background and Literature Review

2.1. Edwards Aquifer

The EA is a crucial water source for municipal, industrial, and agricultural pumping users and supports the springflow needs of endangered species in south-central Texas. Balancing water allocation between pumping users and springflows has been debated for over two decades. EA recharge depends on rainfall and is highly variable. Recharge has varied widely: in 1956, it was 43.7 thousand acre-feet with precipitation of 11.22 inches; and in 1992, the recharge reached 2,176.1 thousand acre-feet with 38.31 inches. Significant droughts in the 1950s resulted in the cessation of flows in the Comal Springs, which led to the extinction of the fountain darter population [14]. Habitat for several endangered species is supported by springflows [15]. The aquifer is karstic and does not retain water as it fell greatly in the year after the large recharge events [16].

In Texas, surface water is governed by the prior appropriation doctrine while groundwater is governed by the rule of capture, which discourages the conjunctive use of groundwater and surface water. The groundwater use can be regulated by the Texas Supreme Court with that regulation subjected to compensation [14]. In early 1993, springflow protection was ordered by a federal court based on a suit under the endangered species act. Then, the Texas legislature passed Texas Senate Bill 1477 (SB1477), which created the Edwards Aquifer Authority (EAA) and directed the EAA to manage the aquifer withdrawals. SB1477 required the maximum annual volume of water pumped to be 400 thousand acre-feet by 2008 and mandated establishment of water rights. Furthermore, regional water managers, in order to protect endangered species, introduced springflow and aquifer elevation-dependent management. The Critical Period Management Plan (CPMP) is currently used for management and makes pumping dependent on aquifer elevation and springflow. For example, the CPMP requires a permitted withdrawal reduction of 20% when the 10-day average of the rate of flow of the Comal Springs is below 225 cubic feet per second (cfs).

2.2. Literature Review

The US southwest is projected to have more frequent multi-year droughts [17,18] and reduced cool season precipitation [19]. Diffenbaugh et al. [20] concluded that global warming was increasing drought probability. Sheffield and Wood [9] projected that central North American long-term droughts would become three times more common. Such developments stress the regional water situation and enhance water competition.

Conjunctive use of ground and surface water is a common strategy for managing drought in arid regions [21,22]. Daneshmand et al. [23] applied an integrated hydrologic, socio-economic, and environmental approach to assess conjunctive water use during drought in the Zayandehrood water basin in Iran, and found that conjunctive use would preserve water supply reductions to under 10% of the irrigation demand during a drought. Pulido-Velazquez et al. [24] analyzed the economic and reliability benefits from different conjunctive uses of surface and groundwater in southern California and noted that conjunctive operations could be adjusted in anticipation of drought and wet years to reduce water scarcity and scarcity cost.

On an aquifer scale, Castaño et al. [25] used a groundwater flow model to evaluate the impacts of drought cycles (from 1980 to 2008) on the evolution of groundwater reserves in the Mancha oriental aquifer system (SE Spain), and their results showed that if the drought was to persist, the costs from the storage deficit ranged from €21.7 million to €34.9 million. Golden and Johnson [26] developed an economic model of production and temporal allocation to estimate producer and hydrologic impacts over a 60-year time horizon in the Ogallala aquifer area in northwest Kansas, and found that limited irrigation was the least costly method of conserving water. The importance of groundwater management under drought conditions has also been found along the Changjiang River, China [27], South Africa [28], and northwestern Bangladesh [29].

Several studies have been conducted on the EA. Scanlon et al. [30] used the equivalent porous media models to simulate groundwater flow in Barton Springs Edwards aquifer. Loaiciga et al. [31] scaled historical recharge data to 2×CO_2 conditions to set up recharge scenario and found that the water resources of the EA could be severely impacted under warmer climate scenario if aquifer recharge and pumping strategies were not properly considered. McCarl et al. [32] used the Edwards Aquifer Simulation Model (EDSIM) to examine the economic dimensions of water management policy on the EA region. EDSIM is an economic and hydrological simulation model that depicts water allocation, agriculture, municipal/industrial (M&I) use, springflow and pumping lifts in the EA, and it depicts the water supply and use across nine states defined by the probability distribution of recharge. These states represent the full spectrum of recharge possibilities, and lower recharge years are used in this study to represent drought. The drought defined here is meteorological since recharge in a karst aquifer correlates well with rainfall variation [16]. Subsequently, EDSIM has been used to study climate change effects [33], regional water planning [34], El Niño-Southern Oscillation (ENSO) effects [12], and elevation dependent management [35]. To date, EA studies have not considered the possible discontinuation of cropping with the conversion of agricultural land to livestock pasture. Use of integrated crop-livestock systems represents a method of adapting to increased drought [36,37]. A number of studies have been performed on the economic impacts of increased drought occurrence, with most studies focused on surface water [38–40].

To conduct our research, we modified the EDSIM model used by McCarl and team [12,32–35] adding livestock production, land conversion to grassland from irrigated cropping in turn supporting livestock production and used it under scenarios exhibiting an increased probability of drought occurrences.

3. Modeling Framework

The main model used here is the EDSIM, which simulates the agricultural and municipal and industrial (M&I) water uses, compares irrigated versus dryland cropping, and considers the livestock herd size, pumping cost and springflow. This model optimizes the consumers' and producers' surpluses by simulating the economic allocation of land and water in a perfectly competitive economy (as discussed in McCarl and Spreen [41] and Lambert et al. [42]) subject to legislatively imposed pumping limits.

EDSIM is a two-stage stochastic model [43] with nine states of nature representing different recharge amounts and climate conditions. At the first stage, the choice of developed irrigated land, land conversion between irrigated land and grassland and dryland, crop mix and livestock herd size is decided independently of recharge state. At the second stage, the recharge state is taken into account. The crop irrigation strategy, crop harvesting, livestock feeding, and M&I water use can be adjusted when the recharge state is known. The irrigation strategy is decided with knowledge of the recharge state, yield consequences, pumping lift, and crop mix. Livestock production is not directly affected by the water availability. Livestock competes with crops as more livestock land arises only through land conversion between cropland and grassland. Water use in the M&I sectors is dependent on the recharge state plus pumping lift. Pumping lift is a function of aquifer elevation which in turn is a function of initial aquifer level, recharge amount and pumping use. Two aquifer pools are modeled one in the east and one in the west. The volume of springflow is determined by initial aquifer level, recharge amount and pumping by agricultural and non-agricultural sectors.

A mathematical presentation on the model is given in Appendix A.

4. Scenario Setup

The analysis on the effects of increasing drought was conducted by running the model under alternative scenarios. These scenarios contained changes in the probability of drought occurrence plus changes with and without pumping and springflow limits as well as at different times considering population growth. The specific scenarios are defined in Table 1.

- Under the increased drought frequency scenarios, the probability of drought events with lower recharge level in the 78-year (year 1934–2011) distribution was increased. According to the recent work by Aryal et al. [8], a 20% increase in drought frequency is projected. Hence, we followed Adamson et al. [38] and increased the probability of drought years so they were some 20% larger while decreasing the probability of the rest of the years so it was some 20% smaller, with the probability of normal years unchanged.
- A maximum pumping limit of 400 thousand acre-feet was considered based on SB1477. Another scenario of a minimum springflow of 225 cfs was introduced to take into account endangered species protection. This is essentially a strategy currently being utilized in the region. Additionally, we examined a maximum lower pumping limit of 375 thousand acre-feet to investigate how it performed under the increased drought.
- We considered M&I demand growth stimulated by population growth in the form of a 10% increase in water demand by the M&I sectors.

Table 1. Definition of scenarios.

Scenarios	Definition
2011Base	Baseline
2011Base400	Base model with pumping limit of 400 thousand acre-feet
2011Base375	Base model with pumping limit of 375 thousand acre-feet
2011Base+Spring225	Base model with minimum springflow of 225 cfs
10Base400	M&I water demand increases of 10% and 400 thousand acre-feet pumping limit
10Base375	M&I water demand increases of 10% and 375 thousand acre-feet pumping limit
10Base+Spring225	M&I water demand increases of 10% and minimum springflow of 225 cfs

5. Data Specification

EDSIM depicts the activity in parts of six counties that constitute the recharge and pumping use zone of the EA. The counties are Kinney, Uvalde, Medina, Bexar, Comal, and Hays. Study area is shown in Figure 1. Data are generally at the county level. When county-level data were unavailable, then district data were used mainly for crop and livestock production budget data. We discussed the basics of the data below with details provided in Ding [44].

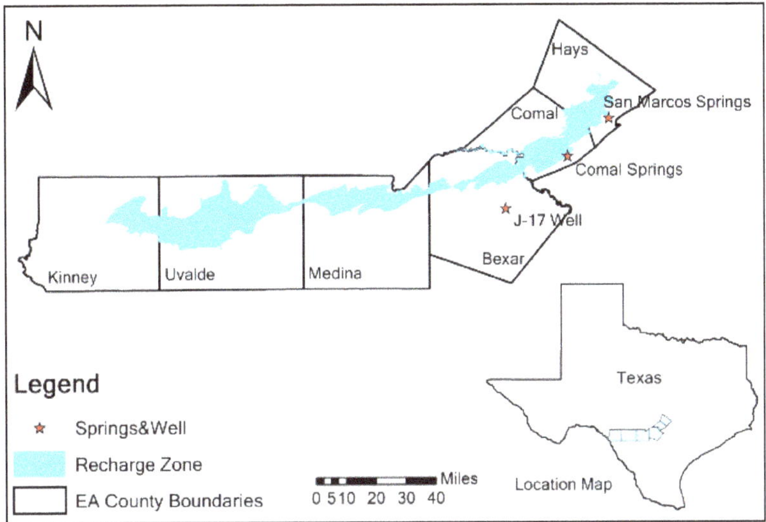

Figure 1. Edwards aquifer region and typical springs and well.

5.1. Crop and Livestock Data

Crop budget data were drawn from the annual budgets produced by the Texas A&M AgriLife Extension Service. These budget data include crop yield, price, and input cost with year 2011 being used. Crop mix data were drawn from Quick Stats, National Agricultural Statistics Services (NASS) and the Census of Agriculture. The mix data included the harvested acreage by crop.

The livestock budgets also came from the Texas A&M Agrilife Extension Service although due to data availability we had used information from an adjacent region [45]. The budgets were defined on an animal unit (AU) basis. One head of cattle was treated as one animal unit, and six head of goats and five head of sheep were also considered on one animal unit [46]. The net benefit per AU was specified as the returns above direct expenses less the cost of grassland use per acre per AU. Livestock mix were defined based on inventory which were collected from Quick Stats, NASS.

5.2. Recharge Data and States of Nature

Following the original EDSIM, there are nine recharge states that range from heavily dry to heavily wet based on regional US geological survey data as reported by the EAA. To form these, we clustered the historical annual recharge data into the nine states of nature. Table 2 shows the recharge states and the corresponding typical weather years clustered into each. The probability of a recharge state was defined as the relative incidence of a weather years falling into that state thus, since 3 years fell into the 'heavily dry' category, we used 3 divided by 78 as the probability. Based on the typical weather years, we obtained the probability distribution of the recharge state (see Table 2). Following Cai [47], the dry years in Table 2 were classified as drought years; normal as normal years, and the remainder as wet years. Hence, the probabilities for drought, normal, and wet years were 0.1923, 0.4615, and 0.3462, respectively. In the scenario of increased drought frequency, the summed probability of drought years increased from 0.1923 to 0.3923 by essentially doubling those probabilities and the summed probability of wet years were decreased from 0.3462 to 0.1362. The probability of normal years was not changed. The drought frequency is simply defined as the summed probability of drought years without considering the temporal or seasonal nature of droughts.

Table 2. Recharge states, years represented, recharge level, and probability distribution.

Recharge State	Years (1934-2011) (Typical Weather Years in Bold)	Recharge Level (10^3 acre-feet)	Probability
Heavily dry	**1956**, 2011, 1951	43.7	0.0385
Medium dry	1954, 1953, **1963**, 1948, 1934	170.7	0.0641
Dry	1955, 1984, 1950, 2006, 2008, 2009, **1989**	214.4	0.0897
Dry-normal	1962, 1943, **1952**, 1940	275.5	0.0513
Normal	**1996**, 1988, 1939, 1937, 1980, 1964, 1983, 1982, 1947, 1938, 1993, 1967, 1999, 1978, 1949	324.3	0.1923
Normal-wet	1945, 1995, 1994, 1946, 1942, 1944, 1969, 2000, 1966, 1965, **1974**, 1970, 2003, 1959, 1961, 2005, 1972	658.5	0.2179
Wet	2010, 1960, 1941, 1968, **1976**, 1936, 1971, 1977, 1975, 1985, 2001, 1979, 1990, 1997, 1998, 1957, 1986	894.1	0.2179
Medium wet	**1935**, 1981, 1973, 1991, 2002, 1958	1711.2	0.0769
Heavily wet	**1987**, 2004, 2007, 1992	2003.6	0.0513
Average		710.9	

5.3. Land Availability

Land availability data were obtained from the 2007 Census of Agriculture. Cropland was categorized as irrigated land and dryland, and irrigated land was further classified as furrow and sprinkler land. Three pumping lift zones were considered here to reflect initial depth to water. The availability of sprinkler land in each lift zone was calculated based on the zonal percentage of total pumping use, and then the available furrow land in each zone was the difference between the irrigated land in each zone and the estimated sprinkler land. As in McCarl et al. [32], dryland was initially set as zero because we focused on studying land use and conversion.

Grassland use was added to the EDSIM. We assumed that all of the grassland is non-irrigated [45]. Furrow or sprinkler land can be converted to dryland or grassland.

5.4. Municipal and Industrial Water Usage

Water usage data in the M&I sectors were based on the Hydrologic Data Report from the EAA website [48]. The water usage data were annual, although we needed monthly data for the EDSIM. Therefore, the 2011 monthly M&I water usage data were calculated based on the monthly distribution of water use in 1996. The elasticity of municipal demand was from Griffin and Chang [49], while the elasticity of industrial demand came from Renzetti [50].

5.5. Linkages between Water Usage, Spring Flows, and Aquifer Elevation

A critical part of the study involved linking the spring flow and aquifer elevation to water usage. This was done using functions estimated by Keplinger and McCarl [51] that related spring flow and ending elevation levels to the initial water level, pumping, and recharge level by states of nature.

6. Model Results and Discussion

In doing our analysis, we first solved the model with and without increasing the drought frequency and reported the results on welfare, land use, water use, springflow, and ending elevation under various scenarios. Later we examined how welfare changes under different degrees of increased drought incidence.

6.1. Welfare Effects

Table 3 presents the welfare effects with and without increased drought frequency. First, we considered the base results for no changes in drought probability. Under 2011 conditions, the crop income is $211.75 million and livestock income is $54.80 million. When a 400 thousand acre-feet limit (2011Base400) is considered, the results show a crop income reduction of $8.24 million per year, which is 3.89% below the baseline income level. Income from livestock production increases by $2.38 million, which is 4.34% of the base year income. This reflects land moving out of irrigation and into grassland. The loss in M&I surplus is less than 0.1% of the baseline surplus. The percentage change in M&I surplus is small because the water demand curve in these two sectors is fairly inelastic with water values being substantially higher.

Table 3. Comparison of the welfare effect with and without increasing drought frequency.

Scenarios		Change in Economic Benefit (10^6\$)			
		Cropping	Livestock	M&I	Total Surplus
Prob(Drought) No Change	2011Base_Baseline	211.75	54.80	828.41	1094.95
	2011Base400	−8.24	2.38	−0.64	−6.50
	2011Base375	−11.22	3.12	−0.83	−8.92
	2011Base+Spring225	−4.73	1.44	−0.65	−3.94
	10Base400	−11.47	3.20	82.04	73.72
	10Base375	−14.70	3.97	81.82	71.09
	10Base+Spring225	−8.13	2.24	82.20	76.31
Prob(Drought) Increases 0.2	2011Base	−6.88	0.00	4.48	−2.40
	2011Base400	−14.38	2.35	3.79	−8.25
	2011Base375	−17.20	3.03	3.51	−10.67
	2011Base+Spring225	−13.41	2.10	3.87	−7.44
	10Base400	−17.45	3.09	86.78	72.43
	10Base375	−21.10	4.12	86.80	69.82
	10Base+Spring225	−16.55	2.92	86.98	73.35

Note: Definitions of scenarios are provided in Table 1.

If the pumping limit is stricter, e.g., 375 thousand acre-feet, then the welfare changes in each sector are larger. Compared with the effects under pumping limits of 400 thousand acre-feet and 375 thousand acre-feet, the effects of a springflow limit of 225 cfs on welfare are smaller because the total water use under this limit is greater than 400 thousand acre-feet (see Table 5). The springflow limit is not as strict in limiting the water use in the EA region because the springflow limit allows more water use in wet years and in fact is the way that the aquifer is managed currently.

Moreover, if the M&I water demand increases by 10%, which is observed in the 10Base400, 10Base375 and 10Base+Spring225 scenarios, crop income shows greater decreases while livestock income increases slightly; however, the M&I surplus increases greatly because the water demand curves for these sectors shift outward.

A drought probability increase of 0.2 yields more extreme results. If a pumping limit is not considered, then increased drought will lead to a cropping loss of $6.88 million and a total surplus loss of $2.40 million with water flowing to M&I interests. These losses are larger under pumping limits, and they also significantly reduce springflow, which will be shown in a later hydrologic section. Under a 400 thousand acre-feet pumping limit, additional frequent droughts will cause a greater cropping loss of $6.14 million. Income from the livestock sector decreases slightly, whereas the M&I surplus increases because water flows to more valued users. Increased drought also results in a total welfare loss of $1.75 million per year. When stricter pumping limits are imposed, cropping losses due to more frequent droughts are lower. For example, under 375 thousand acre-feet pumping limit (2011Base375), increased drought will cause cropping losses of $5.98 million, whereas additional frequent droughts under 400 thousand acre-feet pumping limit will cause crop losses of $6.14 million.

As water demand in the M&I sectors increases and lower pumping limits are imposed, then cropping income declines more and livestock income increases. If the M&I water demand increases by 10%, more frequent drought will increase the competition for water allocation among irrigation, municipal, and industrial users. Additional water flows to the M&I sectors leads to more losses in cropping income. For example, under a minimum springflow of 225 cfs (10Base+Spring225), increased drought causes a cropping loss of $8.42 million per year and livestock income increase of $0.68 million per year.

6.2. Land Use

Data in Table 4 portray land use impacts with and without altered drought frequency. There is no irrigated land converted to grassland when no pumping limits are imposed. For cases of no changes in drought incidence, the lower pumping limit of 400 thousand acre-feet results in land conversion of 10.54 thousand acres from furrow land to grassland and 20.65 thousand acres from sprinkler land to grassland. When much stricter pumping limits are imposed (2011Base375), more furrow land and sprinkler land are converted to grassland. The impact from imposing a minimum springflow constraint is smaller than that from a pumping limit for the same reason provided above. The impacts on land use will be greater if there is an increase in M&I water demand of 10%.

Table 4. Comparison of impacts on land conversion with and without increasing drought frequency.

Scenarios		Change in Land Use (10^3 acre-feet)	
		FurrowToGrass	SprinklerToGrass
Prob(Drought) No Change	2011Base_Baseline	0.00	0.00
	2011Base400	10.54	20.65
	2011Base375	15.73	25.45
	2011Base+Spring225	0.00	19.47
	10Base400	15.73	26.59
	10Base375	17.80	35.27
	10Base+Spring225	0.00	30.97
Prob(Drought) Increases 0.2	2011Base	0.00	0.00
	2011Base400	0.00	32.17
	2011Base375	0.00	41.93
	2011Base+Spring225	0.00	28.96
	10Base400	0.77	42.03
	10Base375	13.76	42.03
	10Base+Spring225	0.00	40.56

Note: (1) Definitions of scenarios are provided in Table 1. (2) When the drought frequency is increased by 0.2, each scenario is prefixed with "Prob". (3) FurrowToGrass denotes furrow land converted to grassland. SprinklerToGrass is referred to land conversion from sprinklered land to grassland. There is no land conversion from irrigated land to dryland.

When the drought probability increases by 0.2 under the 2011Base400 scenario, the conversion of furrow land to grassland becomes 0 acres, whereas additional frequent droughts causes more conversion of 11.52 thousand acres from sprinkler land to grassland. Increased land conversion of sprinkler land to grassland also occurs under the other scenarios. For instance, under the scenario of 2011Base375, more frequent droughts reduce the sprinkler land by 16.48 thousand acres via the conversion to grassland. Furthermore, the drought impact on land use change increases in severity when the M&I water demand increases by 10%, which also increases the conversion of irrigated land to grassland. For example, a comparison of scenarios 10Base400 with and without increasing drought frequency shows that when drought becomes more frequent, additional sprinkler land is converted to grassland while less furrow land is converted to grassland. Land transfers increase when water allocation becomes more competitive, i.e., under a pumping limit of 375 thousand acre-feet.

6.3. Water Use

Table 5 shows water use with and without increases in drought frequency. When droughts do not increase and the total water withdrawn from the aquifer is restricted to 400 thousand acre-feet, then the total water usage is reduced by 102.17 thousand acre-feet, with 89.88% of the reduction from agriculture, primarily in the east region (see Figure 2). When springflow is limited to be greater than

225 cfs, east agricultural water use also decreases considerably. A 10% increase in M&I water demand further reduces the water usage by agriculture.

Table 5. Comparison of the impacts on water use with and without increased drought.

Scenarios		Change in Water Use (10^3 acre-feet)		
		Irrigated Cropping	**M&I**	**Total Value**
Prob(Drought) No Change	2011Base_Baseline	224.10	284.90	509.01
	2011Base400	−91.83	−10.34	−102.17
	2011Base375	−114.82	−12.03	−126.85
	2011Base+Spring225	−57.55	−11.04	−68.59
	10Base400	−117.06	15.21	−101.85
	10Base375	−140.71	13.86	−126.85
	10Base+Spring225	−82.90	16.36	−66.55
Prob(Drought) Increases 0.2	2011Base	−0.30	1.49	1.18
	2011Base400	−93.99	−8.74	−102.73
	2011Base375	−116.30	−10.64	−126.95
	2011Base+Spring225	−74.98	−8.90	−83.88
	10Base400	−118.14	16.51	−101.63
	10Base375	−143.37	16.56	−126.81
	10Base+Spring225	−102.36	17.41	−84.95

Note: Definitions of scenarios are provided in Table 1.

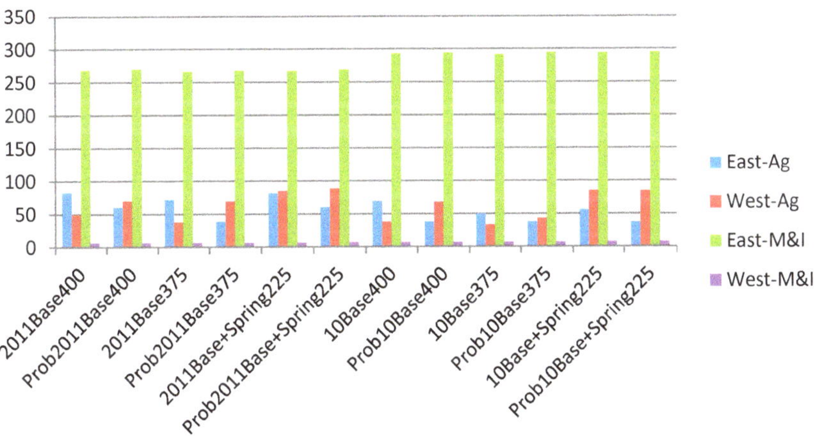

Figure 2. Water use in the east and west region in the agricultural and M&I sectors. Note: (1) Definitions of scenarios are provided in Table 1. (2) When the drought frequency is increased by 0.2, each scenario is prefixed with "Prob". (3) East-Ag and West-Ag represent the water use of agriculture in east and west EA regions, respectively, and East-M&I and West-M&I are the municipal and industrial water use in the east and west EA region, respectively.

Now, we consider the effect of increased drought on water use. When the total water pumping is limited to 400 thousand acre-feet, more frequent droughts will cause a further reduction of agricultural water use of 2,160 acre-feet and a total water usage decrease of 560 acre-feet. If the total water pumped from the aquifer is restricted to 375 thousand acre-feet, then agricultural water use further declines by 1,480 acre-feet and the total water usage is reduced by 100 acre-feet. However, when springflow limit of 225 cfs is imposed, more frequent droughts will cause more reduction in agricultural water use. The above comparison indicates that stricter pumping constraints lower the impact of increased drought on water allocation because ample water is frequently observed. Furthermore, drought impacts on irrigation water use increase if the M&I water demand increases by 10%, with the impacts mainly on eastern region irrigation water use (see Figure 2).

6.4. Hydrologic Impacts

Figure 3 presents a comparison of the hydrologic impacts. When the drought probability is not changed, both the pumping and minimum springflow limits increase the springflow in both Comal Spring and San Marcos Spring, and the J17 well water elevation as well. The lower pumping limit (375 thousand acre-feet) increases the springflow and the J17 well water elevation the most. If the M&I water demand keeps unchanged, the 400 pumping limit protects the springflow and well elevation better than does the springflow restriction. However, when the M&I water demand goes up by 10%, the role of springflow restriction is bigger than 400 pumping limit.

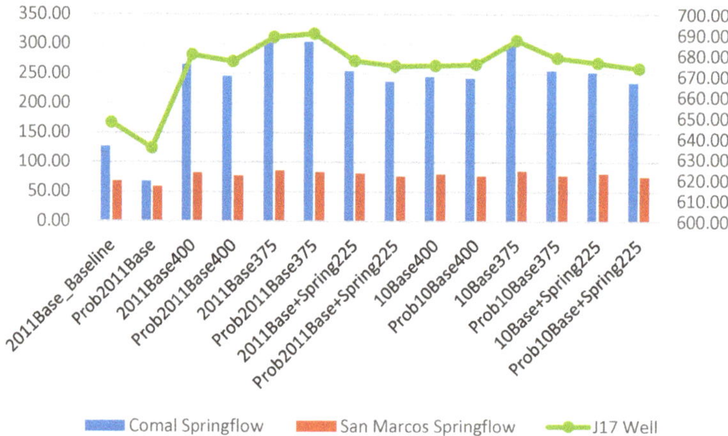

Figure 3. Comparison of hydrologic impacts with and without increased drought. Note: (1) Definitions of scenarios are provided in Table 1. (2) When the drought frequency is increased by 0.2, each scenario is prefixed with "Prob". (3) "Comal Springflow" and "San Marcos Springflow" refer to the springflow in Comal Spring and San Marcos Spring, respectively (10^3 acre feet). (4) "J17 Well" refers to the elevation of a reference well in San Antonio (feet).

When the drought probability increases by 0.2, the springflow and the J17 well elevation are reduced. If there are no restrictions on pumping or springflow, increased drought will cause the springflow and J17 well end elevation decline greatly, which further emphasizes the importance of pumping limits and/or minimum springflow limits. Again, the impacts are smaller with a pumping restriction of 375 thousand acre-feet because this limit provide a safety margin. Similar results are observed when the M&I water demand increases. Note here in the case if both the M&I water demand and drought frequency increase, pumping limits can help better protect springflow and well elevation than springflow restriction, probably because pumping limits can overall plan the use of water resources when water use competition is stricter.

6.5. Comparison of the Impacts under Different Changes in Drought Probability

Table 6 reports the impacts under different changes in drought frequency. Here we consider drought probability increase from 0.1 to 0.3, which holds the probability of normal years unchanged and let the drought probability increase and the wet probability decrease in the relative amount. According to Zhao et al. [52], they projected that the probability distribution function (pdf) of agricultural drought would become flatter. We also consider a scenario that drought probability increases 0.2 and normal year probability and wet probability decreases 0.1, respectively. The first four lines of Table 6 present the average economic benefit under different degrees of drought frequency change. The baseline presents the case when the total water pumped is limited to 400 thousand acre-feet. In turn, if the probability of drought increases by 0.1, then cropping will suffer a loss of $2.97 million. Moreover, the income from livestock production will decrease slightly, the M&I surplus will increase by $2.13 million, and increased drought will cause a total surplus loss of $1.03 million per year.

Table 6. Comparison of impacts with various degrees of drought probability change.

	2011Base400 (Baseline)	Change from Baseline			
		Prob + 0.1	Prob +0.2	Prob + 0.3	Prob + 0.2 & Flattening pdf
Economic Benefit($10^6$$)					
Cropping	203.51	−2.97	−6.14	−9.24	−5.30
Livestock	57.18	−0.19	−0.03	0.06	0.08
M&I	827.76	2.13	4.43	6.80	2.66
Total Surplus	1088.45	−1.03	−1.74	−2.38	−2.56
Land Use (10^3acres)					
Irrigated Land	66.09	0.76	−1.43	−2.80	−1.24
Dryland	0.00	0.00	0.00	0.00	0.00
Grassland	740.26	−1.24	0.98	2.36	1.01
Water Use (10^3acre-feet)					
Irrigated Cropping	132.275	0.07	−2.16	−4.01	−0.37
M&I	274.563	0.22	1.61	3.02	0.55
Total Value	406.838	0.30	−0.56	−0.99	0.18
Hydrologic Effects					
Comal Spring flow (10^3acre-feet)	265.32	5.37	−20.15	−47.93	−45.12
San Marcos Spring flow (10^3acre-feet)	81.40	−0.88	−5.04	−9.43	−6.51
J-17 Well End Elevation (feet)	680.41	1.90	−3.32	−9.03	−9.72

Note: (1) 2011Base400 is the baseline. (2) Prob(Drought) refers to the probability of drought. "Prob(Drought) increases 0.2" means the probability of drought years increases by 0.2, while the probability of wet years decreases by 0.2. (3) "Prob(Drought) increases 0.2 with flattening pdf" means that the probability of drought years increases by 0.2, while the probability of normal years and the probability of wet years decreases by 0.1, respectively.

When droughts become more frequent, the cropping loss will be greater and the acreage of irrigated land will decrease. When the drought probability increases by 0.1, additional irrigated farming is conducted; however, as droughts become more frequent, irrigated acreage decreases and more land is converted to grassland. When grassland acreage increases, livestock income increases as well. In terms of water use and hydrologic impact, water reductions primarily occur in the irrigation use. Additionally, as droughts become more frequent, the springflow in both of the springs and the J17 well water elevation are reduced. More frequent droughts reduce the springflow, and stricter pumping limits or springflow restrictions would be required to maintain current springflow levels and protect spring-supported endangered species.

7. Conclusions

EA recharge mainly relies on rainfall, which would be negatively affected by the increased incidence of drought proved by a number of studies in the IPCC [2]. In particular, drought frequency is predicted to increase in the southwestern U.S., where the EA is located [17,18]. We find that such developments would shift water from cropping to M&I interests with decreases in cropping income, increases in livestock income and not much effect on the M&I welfare. We find that under a pumping limit of 400 thousand acre-feet, an increased drought frequency will result in a regional cropping loss of $6.14 million per year, with yet more water reallocated to M&I interests. Stricter pumping limitations, such as a 375 thousand acre-feet pumping limit, help alleviate the cropping losses under the increased drought scenario principally due to lower pump lifts.

We also found that more frequent droughts will increase land transfers from irrigated land to grassland and livestock uses, while decreasing springflows in both Comal Spring and San Marcos Spring. To preserve the endangered species habitat surrounding the springs, lower maximum pumping limits or minimum springflow restrictions are required. There are also ecological implications of this in that there are several endangered species whose habitat is supplied by the springflow plus a change from irrigated agriculture to grassland-based livestock would certainly bring about a number of other ecological alterations.

Author Contributions: Conceptualization, B.A.M.; methodology, B.A.M. and J.D.; software, J.D.; validation, J.D.; formal analysis, J.D.; investigation, J.D.; resources, B.A.M. and J.D.; data curation, J.D.; writing—original draft preparation, J.D.; writing—review and editing, B.A.M. and J.D.; visualization, J.D.; supervision, B.A.M.; project administration, J.D.; funding acquisition, B.A.M. and J.D. All authors have read and agreed to the published version of the manuscript.

Funding: This research was partially funded by the NOAA—Climate Programs Office—Sectoral Applications Research Program under grant NA12OAR4310097 and the U.S. Department of Agriculture—National Institute of Food and Agriculture under grant 2011-67003-30213 in the NSF—USDA—DOE Earth System Modelling Program. In addition while the majority of this work was done at Texas A&M University, some was also done at Xiamen University with funding provided by the Chinese Fundamental Research Funds for the Central Universities under grant 20720151281. We acknowledge the support of both institutions.

Acknowledgments: We thank David Anderson and Chengcheng Fei at Texas A&M University for their helpful comments. Two anonymous referees provided comments that improved the initial version of this paper.

Conflicts of Interest: The authors declare no conflict of interest.

Appendix A EDSIM Model Concept and Structure

Before presenting the fundamental algebraic structure of the EDSIM, we will overview its theoretical structure. In brief, EDSIM is a price endogenous mathematical program that can be represented by the following Equation (A1),

$$
\begin{aligned}
Max : \int_0^{Q_d} P_d(Q_d)dQ_d - \int_0^{Q_s} P_s(Q_s)dQ_s \\
s.t.\ Q_d - Q_s \le 0 \\
Q_d, Q_s \ge 0
\end{aligned}
\tag{A1}
$$

where Q_d and Q_s are the quantities demanded and supplied, respectively; $P_d(Q_d)$ is the inverse demand curve that provides the demand price as a function of the quantity demanded; and $P_s(Q_s)$ is the inverse supply curve that provides the supply price as a function of the quantity supplied. The objective function is the sum of the consumers' plus producers' surplus in the EA region subject to hydrological, land, and institutional constraints. The model is stochastic facing nine states of nature where the model must make certain choices (crop mix, livestock herd size, land transformation) before knowing the state of nature and then others dependent on the state of nature (irrigation strategy, crop harvest and sale, livestock sale, municipal and industrial water consumption). The first-order conditions of such a model render the model solution as a simulation of what would happen under a perfectly competitive regional allocation of resources [41,42].

Appendix A.1 Objective Function

The objective function depicts the expected consumers' plus producers' surpluses. In this case due to the extremely small share of US production in the region fixed prices are used for agricultural commodities. Consequently, the objective function maximizes the revenue from crop and livestock production plus the area under the M&I demand curves less the costs of crop production, livestock production, and development of irrigated land and costs of lift-dependent pumping. Also, stochastics are present with some terms independent of state of nature and other terms dependent. More precisely, the objective function is presented as follows, with variables in upper case and parameters in lower case.

The first part (first line) of Equation (A2) contains two costs and is independent of state of nature. One is the unit cost of irrigation development (*irrcost*) by lift zone (*z*) multiplied by the irrigated land developed (*IRRLAND*), the other is the cost of converting furrow land to sprinkler land (*FURRTOSPK*) in a county (*p*) and lift zone (*z*).

$$Max: \; -\sum_{p}\sum_{z} irrcost_z IRRLAND_{pz} - \sum_{p}\sum_{z} sprinkcost_p FURRTOSPK_{pz}$$

$$+\sum_{r} prob_r \left(\begin{array}{c} +\sum_{p}\sum_{z}\sum_{c}\sum_{s} irrincome_{rcs} IRRPROD_{pzrcs} \\[4pt] +\sum_{p}\sum_{c} dryincome_{rc} DRYPROD_{prc} \\[4pt] -\sum_{p}\sum_{z}\sum_{m} AGPUMPCOST_{pzr} AGWATER_{pzrm} \\[4pt] +\sum_{p}\sum_{z}\sum_{l} liveincome_{rl} LIVEPROD_{pzrl} \\[4pt] -\sum_{p}\sum_{z} grasscost_r GRASSUSE_{pzr} \\[4pt] +\sum_{p}\sum_{m} \int_{0}^{MUN_{prm}} mp_{prm}(MUN_{prm}) dMUN_{prm} \\[4pt] +\sum_{p}\sum_{m} \int_{0}^{IND_{prm}} ip_{prm}(IND_{prm}) dIND_{prm} \\[4pt] -\sum_{p}\sum_{m} MIPUMPCOST_{pr}(MUN_{prm} + IND_{prm}) \end{array} \right) \qquad (A2)$$

The second part of Equation (A2) in brackets is stochastic based on the recharge state (*r*) and is weighted by the probability (*prob*) of each state. The first two lines depict the net revenue from crop yields, which is the crop revenue minus production costs per acre (*irrincome* and *dryincome*) multiplied by the acres produced (*IRRPROD* and *DRYPROD*) summed across each county (*p*), pumping zone (*z*), crop (*c*), recharge state (*r*), and irrigation strategy (*s*). The third line subtracts irrigation water pumping cost (the variable *AGPUMPCOST*) multiplied by water use (*AGWATER*) by county (*p*) and lift zone (*z*) in month (*m*) under recharge state (*r*). Lines 4 and 5 represent livestock production net revenue, which includes the livestock net income (*liveincome*) multiplied by livestock raised (*LIVEPROD*) by livestock type (*l*), county (*p*) and lift zone (*z*) under the recharge state (*r*). We also deducted the cost of grassland maintenance (*grasscost*) times the grassland used (*GRASSUSE*) by county (*p*) and lift zone (*z*) and recharge state (*r*). The last three lines represent the M&I benefits and costs of water pumping, which involves the area under the M&I demand curves less pumping cost (a variable *MIPUMPCOST*) by county (*p*) under recharge state (*r*). The variables *MUN* and *IND* represent the amount of water demanded in the M&I sectors, respectively.

Appendix A.2 Land Availability Constraint

Equation (A3) limits irrigated land use by crop (*c*) and irrigation strategy (*s*) in a lift zone (*z*) and county (*p*) under recharge state (*r*) (*IRRPROD*) to the total irrigated land (*IRRLAND*). The total irrigated land does not vary by recharge state, meaning that it is set before climate conditions are known, however, the irrigated land choice is in a recharge state dependent on the crop use and irrigation strategy.

$$\sum_{c}\sum_{s} IRRPROD_{pzrcs} - IRRLAND_{pz} = 0 \; for \; all \; p, z, r \qquad (A3)$$

The initial availability of dryland is zero because we only examine initially irrigated land area. Equation (A4) requires that dryland use in a county (*DRYPROD*) does not exceed the land converted from irrigated land to dryland (*IRRTODRY*) by county and lift zone. Note that the dryland available through conversion is the same across all recharge states, but the dryland use can vary by recharge state.

$$\sum_{c} DRYPROD_{prc} - \sum_{z} IRRTODRY_{pz} \leq 0 \; for \; all \; p, r \qquad (A4)$$

Equation (A5) balances the total initial irrigated land where land used in irrigated cropping (*IRRLAND*) plus the amount converted to dryland (*IRRTODRY*) or grassland (*IRRTOGRS*) cannot exceed the initial availability (*irrlandavail*) in a county and lift zone. Note that the available converted land is the same across all recharge states.

$$IRRLAND_{pz} - irrlandavail_{pz} + IRRTODRY_{pz} + IRRTOGRS_{pz} \leq 0 \, for \, all \, p, z \qquad (A5)$$

Equation (A6) is the grassland availability constraint, which limits grassland use (*GRASSUSE*) to the initial grassland availability (*grasslandavail*) plus land transformed from irrigated land to grassland (*IRRTOGRS*) by county (p) and lift zone (z). Note that the available grassland is the same across all recharge states but the grassland use can vary by recharge state. Equation (A7) restricts livestock production and grassland use by county (p) and lift zone (z) under recharge state (r), where *gr* denotes the grazing rate, which is the amount of grassland required per animal unit.

$$GRASSUSE_{pzr} - grassavail_{pz} - IRRTOGRS_{pz} \leq 0 \, for \, all \, p, z, r \qquad (A6)$$

$$LIVEPROD_{pzrl} \leq GRASSUSE_{pzr} / gr \, for \, all \, p, z, r, l \qquad (A7)$$

Appendix A.3 Crop Mix Constraint

Following McCarl (1982), the crop mix restriction requires that crop production is a convex combination of historical crop mixes, which is performed for irrigated land and dryland separately. Thus, irrigated land produced (*IRRPROD*) is a convex combination of historical irrigated crop mixes (*irrmixdata*) in terms of crops (c) and mix possibilities (x) in county (p) in Equation (A8). Similarly, dryland produced (*DRYPROD*) is a convex combination of historical dryland crop mixes (*drymixdata*) in Equation (A9). The separate limits for irrigated land and dryland allow their acreage to vary independently as more or less land is converted. A separate mix is allowed for each lift zone causing realistic crop mixes on each zone. The crop mix approach is used to make realistic crop mixes without modeling detailed resource allocation at the farm level [53].

$$\sum_s IRRPROD_{pzrcs} - \sum_x irrmixdata_{pcx} IRRMIX_{px} = 0 \, for \, all \, p, z, r, c \qquad (A8)$$

$$DRYPROD_{prc} - \sum_x drymixdata_{pcx} DRYMIX_{px} = 0 \, for \, all \, p, r, c \qquad (A9)$$

Appendix A.4 Livestock Mix Restriction

Livestock mixes are also defined in Equation (A10). Livestock production (*LIVEPROD*) by livestock type for a county and zone under a recharge state is set to be a convex combination of historical observable livestock mixes (*livemixdata*) in terms of species. As argued by McCarl [53], this constraint can make realistic livestock mixes without modeling the detailed resource allocation at the farm level.

$$LIVEPROD_{pzrl} - \sum_x livemixdata_{plx} LIVEMIX_{px} = 0 \, for \, all \, p, z, r, l \qquad (A10)$$

Appendix A.5 Lift Dependent Pumping Cost

Equations (A11) and (A12) relate the pumping cost to aquifer lift which is determined by the next equation. The agricultural pumping cost per acre-foot of water for county, zone, and recharge state equals a fixed pumping cost (*agcpump*) plus a variable pumping cost (*agvpump*) per foot of lift multiplied by the agricultural lift (*AGLIFT*). Similarly, the M&I pumping costs per acre-foot are similarly defined.

$$AGPUMPCOST_{pzr} = agcpump + agvpumpAGLIFT_{pzr} \, for \, all \, p, z, r \qquad (A11)$$

$$MIPUMPCOST_{pr} = micpump + mivpumpMILIFT_{pr} \, for \, all \, p, r \qquad (A12)$$

Appendix A.6 Aquifer Elevation Determination

The ending water elevation level of the EA is calculated via Equation (A13), which relates the ending water level to a regression-estimated function that was developed by Keplinger and McCarl [51]. Namely we determine elevation as a function of monthly recharge level (*rech*), initial water level (*INITWATER*), and total water use. Total water usage is the sum of municipal (*MUN*), industrial (*IND*), and agricultural (*AGWATER*) water use.

$$
\begin{aligned}
ENDWATER_{wr} - rendint_w - \sum_m rendr_w rech_{rm} - \sum_{w2} rende_{ww2} INITWATER \\
- \sum_{w2} rendu_{ww2} \sum_{p \in reg(w2)} \sum_m (MUN_{prm} + IND_{prm} + \sum_z AGWATER_{pzrm}) = 0 \, for \, all \, w, r
\end{aligned}
\tag{A13}
$$

In Equation (A13), *rendint* is the estimated intercept, *rendr* is the parameter of recharge, *rende* is the initial water parameter, and *rendu* is the parameter of total water use. The subscript w refers to the region where the elevation is calculated, and the subscript $w2$ is used to sum the water use across both the east and the west EA regions.

Appendix A.7 Springflow Equation

Springflow levels are defined in Equation (A14), which relates springflow level to a regression-estimated function again from Keplinger and McCarl [51]. That function relates monthly springflow by spring s to monthly recharge in previous part of year, initial water level, and total water use. Therein *rsprnint* is the estimated intercept, *rsprnr* is the parameter of recharge, *rsprne* is the initial water parameter, and *rsprnu* is the parameter of total water use. Both subscripts m and m^* refer to the month.

$$
\begin{aligned}
SPRINGFLOW_{jrm} - rsprnint_{jm} - \sum_{m^* \leq m} rsprnr_{smm^*} rech_{rm^*} - \sum_w rsprne_{smw} INITWATER_w \\
- \sum_w \sum_{p \in reg(w)} \sum_{m^* \leq m} rsprnu_{smwm^*} * \left(MUN_{prm} + IND_{prm} + \sum_z AGWATER_{pzrm} \right) = 0 \, for \, all \, s, r, m
\end{aligned}
\tag{A14}
$$

References

1. EAA. 2018 Groundwater Discharge and Usage. Available online: https://www.edwardsaquifer.org/wp-content/uploads/2019/08/2018-Discharge-Report.pdf (accessed on 19 December 2019).
2. IPCC. Managing the Risks of Extreme Events and Disasters to Advance Climate Change Adaptation. In *A Special Report of Working Groups I and II of the Intergovernmental Panel on Climate Change*; Field, C.B., Barros, V., Stocker, T.F., Qin, D., Dokken, D.J., Ebi, K.L., Mastrandrea, M.D., Mach, K.J., Plattner, G.-K., Allen, S.K., et al., Eds.; Cambridge University Press: Cambridge, UK; New York, NY, USA, 2012; pp. 25–231.
3. Schwalm, C.R.; Anderegg, W.R.L.; Michalak, A.M.; Fisher, J.B.; Biondi, F.; Koch, G.; Litvak, M.; Ogle, K.; Shaw, J.D.; Wolf, A.; et al. Global Patterns of Drought Recovery. *Nature* **2017**, *548*, 202–205. [CrossRef] [PubMed]
4. Spinoni, J.; Vogt, J.V.; Naumann, G.; Barbosa, P.; Dosio, A. Will Drought Events Become More Frequent and Severe in Europe? *Int. J. Climatol.* **2018**, *38*, 1718–1736. [CrossRef]
5. Touma, D.; Ashfaq, M.; Nayak, M.A.; Kao, S.C.; Diffenbaugh, N.S. A Multi-model and Multi-index Evaluation of Drought Characteristics in the 21st Century. *J. Hydrol.* **2015**, *526*, 196–207. [CrossRef]
6. Yuan, X.; Zhang, M.; Wang, L.; Zhou, T. Understanding and Seasonal Forecasting of Hydrological Drought in the Anthropocene. *Hydrol. Earth Syst. Sci.* **2017**, *21*, 5477–5492. [CrossRef]
7. IPCC. Climate Change 2014: Impacts, Adaptation, and Vulnerability. Part A: Global and Sectoral Aspects. In *Contribution of Working Group II to the Fifth Assessment Report of the Intergovernmental Panel on Climate Change*; Field, C.B., Barros, V.R., Dokken, D.J., Mach, K.J., Mastrandrea, M.D., Bilir, T.E., Chatterjee, M., Ebi, K.L., Estrada, Y.O., Genova, R.C., et al., Eds.; Cambridge University Press: Cambridge, UK; New York, NY, USA, 2014; pp. 229–269.

8. Aryal, Y.; Zhu, J. On Bias Correction in Drought Frequency Analysis Based on Climate Models. *Clim. Chang.* **2017**, *140*, 361–374. [CrossRef]

9. Sheffield, J.; Wood, E.F. Projected Changes in Drought Occurrence under Future Global Warming from Multi-model, Multi-scenario, IPCC AR4 Simulations. *Clim. Dyn.* **2008**, *31*, 79–105. [CrossRef]

10. Trenberth, K.E.; Fasullo, J.T. An Apparent Hiatus in Global Warming? *Earths Future* **2013**, *1*, 19–32. [CrossRef]

11. Bastos, A.; Running, S.W.; Gouveia, C.; Trigo, R.M. The Global NPP Dependence on ENSO: La Niña and the Extraordinary Year of 2011. *J. Geophys. Res. Biogeosci.* **2013**, *118*, 1247–1255. [CrossRef]

12. Chen, C.C.; Gillig, D.; McCarl, B.A.; Williams, R.L. ENSO Impacts on Regional Water Management: Case Study of the Edwards Aquifer (Texas, USA). *Clim. Res.* **2005**, *28*, 175–182. [CrossRef]

13. Long, D.; Scanlon, B.R.; Longuevergne, L.; Sun, A.Y.; Fernando, D.N.; Save, H. GRACE Satellite Monitoring of Large Depletion in Water Storage in Response to the 2011 Drought in Texas. *Geophys. Res. Lett.* **2013**, *40*, 3395–3401. [CrossRef]

14. Gulley, R.L.; Cantwell, J.B. The Edwards Aquifer Water Wars: The Final Chapter? *Tex. Water J.* **2013**, *4*, 1–21.

15. Edwards Aquifer Habitat Conservation Plan 2018 Annual Report. Available online: https://www.edwardsaquifer.org/wp-content/uploads/2019/10/EAHCP_Annual_Report_2018.pdf (accessed on 19 December 2019).

16. Uddameri, V.; Singaraju, S.; Hernandez, E.A. Is Standardized Precipitation Index (SPI) a Useful Indicator to Forecast Groundwater Droughts?—Insights from a Karst Aquifer. *J. Am. Water Resour. Assoc.* **2019**, *55*, 70–88. [CrossRef]

17. Seager, R.; Ting, M.; Held, I.; Kushnir, Y.; Lu, J.; Vecchi, G.; Huang, H.P.; Harnik, N.; Leetmaa, A.; Lau, N.C.; et al. Model Projections of an Imminent Transition to a More Arid Climate in Southwestern North America. *Science* **2007**, *316*, 1181–1184. [CrossRef] [PubMed]

18. Cayan, D.R.; Das, T.; Pierce, D.W.; Barnett, T.P.; Tyree, M.; Gershunov, A. Future Dryness in the Southwest US and the Hydrology of the Early 21st Century Drought. *Proc. Natl. Acad. Sci. USA* **2010**, *107*, 21271–21276. [CrossRef] [PubMed]

19. IPCC. Climate Change 2007: The Physical Science Basis. In *Contribution of Working Group I to the Fourth Assessment Report of the Intergovernmental Panel on Climate Change*; Solomon, S., Qin, D., Manning, M., Chen, Z., Marquis, M., Averyt, K.B., Tignor, M., Miller, H.L., Eds.; Cambridge University Press: Cambridge, UK, 2007; pp. 847–940.

20. Diffenbaugh, N.S.; Swain, D.L.; Touma, D. Anthropogenic Warming Has Increased Drought Risk in California. *Proc. Natl. Acad. Sci. USA* **2015**, *112*, 3931–3936. [CrossRef] [PubMed]

21. Bazargan-Lari, M.R.; Kerachian, R.; Mansoori, A. A Conflict-resolution Model for the Conjunctive Use of Surface and Groundwater Resources that Considers Water-quality Issues: A Case Study. *Environ. Manag.* **2009**, *43*, 470. [CrossRef]

22. Li, Z.; Quan, J.; Li, X.Y.; Wu, X.C.; Wu, H.W.; Li, Y.T.; Li, G.Y. Establishing a Model of Conjunctive Regulation of Surface Water and Groundwater in the Arid Regions. *Agric. Water Manag.* **2016**, *174*, 30–38. [CrossRef]

23. Daneshmand, F.; Karimi, A.; Nikoo, M.R.; Bazargan-Lari, M.R.; Adamowski, J. Mitigating Socio-economic-environmental Impacts during Drought Periods by Optimizing the Conjunctive Management of Water Resources. *Water Resour. Manag.* **2014**, *28*, 1517–1529. [CrossRef]

24. Pulido-Velazquez, M.; Jenkins, M.W.; Lund, J.R. Economic Values for Conjunctive Use and Water Banking in Southern California. *Water Resour. Res.* **2004**, *40*. [CrossRef]

25. Castaño, S.; Sanz, D.; Gómez-Alday, J.J. Sensitivity of a Groundwater Flow Model to Both Climatic Variations and Management Scenarios in a Semi-arid Region of SE Spain. *Water Resour. Manag.* **2013**, *27*, 2089–2101. [CrossRef]

26. Golden, B.; Johnson, J. Potential Economic Impacts of Water-use Changes in Southwest Kansas. *J. Natl. Resour. Policy Res.* **2013**, *5*, 129–145. [CrossRef]

27. Dai, Z.; Du, J.; Chu, A.; Li, J.; Chen, J.; Zhang, X. Groundwater Discharge to the Changjiang River, China, during the Drought Season of 2006: Effects of the Extreme Drought and the Impoundment of the Three Gorges Dam. *Hydrogeol. J.* **2010**, *18*, 359–369. [CrossRef]

28. Calow, R.C.; Robins, N.S.; Macdonald, A.M.; Macdonald, D.M.J.; Gibbs, B.R.; Orpen, W.R.G.; Mtembezeka, P.; Andrews, A.J.; Appiah, S.O. Groundwater Management in Drought-prone Areas of Africa. *Int. J. Water Resour. Dev.* **1997**, *13*, 241–261. [CrossRef]

29. Shahid, S.; Hazarika, M.K. Groundwater Drought in the Northwestern Districts of Bangladesh. *Water Resour. Manag.* **2010**, *24*, 1989–2006. [CrossRef]

30. Scanlon, B.R.; Mace, R.E.; Barrett, M.E.; Smith, B. Can We Simulate Regional Groundwater Flow in a Karst System Using Equivalent Porous Media Models? Case Study, Barton Springs Edwards aquifer, USA. *J. Hydrol.* **2003**, *276*, 137–158. [CrossRef]

31. Loaiciga, H.A.; Maidment, D.R.; Valdes, J.B. Climate-change Impacts in a Regional Karst Aquifer, Texas, USA. *J. Hydrol.* **2000**, *227*, 173–194. [CrossRef]

32. McCarl, B.A.; Dillon, C.R.; Keplinger, K.O.; Williams, R.L. Limiting Pumping from the Edwards Aquifer: An Economic Investigation of Proposals, Water Markets, and Spring Flow Guarantees. *Water Resour. Res.* **1999**, *35*, 1257–1268. [CrossRef]

33. Chen, C.; Gillig, D.; McCarl, B.A. Effects of Climatic Change on A Water Dependent Regional Economy: A study of the Texas Edwards Aquifer. *Clim. Chang.* **2001**, *49*, 397–409. [CrossRef]

34. Gillig, D.; McCarl, B.A.; Boadu, F. An Economic, Hydrologic, and Environmental Assessment of Water Management Alternative Plans for the South Central Texas Region. *J. Agric. Appl. Econ.* **2001**, *33*, 59–78. [CrossRef]

35. Chen, C.; McCarl, B.A.; Williams, R.L. Elevation Dependent Management of the Edwards Aquifer: Linked Mathematical and Dynamic Programming Approach. *J. Water Resour. Plan. Manag.* **2006**, *132*, 330–340. [CrossRef]

36. Gray, E.; Henninger, N.; Reij, C.; Winterbottom, R.; Agostini, P. *Integrated Landscape Approaches for Africa's Drylands*; The World Bank: Washington, DC, USA, 2016; pp. 61–129.

37. Alary, V.; Messad, S.; Aboul-Naga, A.; Osman, M.A.; Daoud, I.; Bonnet, P.; Juanes, X.; Tourrand, J.F. Livelihood Strategies and the Role of Livestock in the Processes of Adaptation to Drought in the Coastal Zone of Western Desert (Egypt). *Agric. Syst.* **2014**, *128*, 44–54. [CrossRef]

38. Adamson, D.; Mallawaarachchi, T.; Quiggin, J. Declining Inflows and More Frequent Droughts in the Murray Darling Basin: Climate Change, Impacts and Adaptation. *Aust. J. Agric. Resour. Econ.* **2009**, *53*, 345–366. [CrossRef]

39. Cañón, J.; González, J.; Valdés, J. Reservoir Operation and Water Allocation to Mitigate Drought Effects in Crops: A Multilevel Optimization Using the Drought Frequency Index. *J. Water Resour. Plan. Manag.* **2009**, *135*, 458–465. [CrossRef]

40. Ward, F.A.; Hurd, B.H.; Rahmani, T.; Gollehon, N. Economic Impacts of Federal Policy Responses to Drought in the Rio Grande Basin. *Water Resour. Res.* **2006**, *42*. [CrossRef]

41. McCarl, B.A.; Spreen, T.H. Price Endogenous Mathematical Programming as a Tool for Sector Analysis. *Am. J. Agric. Econ.* **1980**, *62*, 87–102. [CrossRef]

42. Lambert, D.K.; McCarl, B.A.; He, Q.; Kaylen, M.S.; Rosenthal, W.; Chang, C.C.; Nayda, W.I. Uncertain Yields in Sectoral Welfare Analysis: An Application to Global Warming. *J. Agric. Appl. Econ.* **1995**, *27*, 423–436. [CrossRef]

43. Dantzig, G.B. Linear Programming under Uncertainty. *Manag. Sci.* **1995**, *1*, 197–206. [CrossRef]

44. Ding, J. Three Essays on Climate Variability, Water and Agricultural Production. Ph.D. Thesis, Texas A&M University, College Station, TX, USA, 2014.

45. Anderson, D.; (Texas A&M University: College Station, TX, USA). Personal communication, 2014.

46. Lyons, R.K.; Machen, R.V. Stocking Rate: The Key Grazing Management Decision. Texas FARMER Collection 2004. Available online: https://core.ac.uk/download/pdf/4274892.pdf (accessed on 8 November 2019).

47. Cai, Y. Water Scarcity, Climate Change, and Water Quality: Three Economic Essays. Ph.D. Thesis, Texas A&M University, College Station, TX, USA, 2009.

48. Hydrologic Data Report. Available online: https://www.edwardsaquifer.org/science-maps/research-scientific-reports/hydrologic-data-reports/ (accessed on 19 December 2019).

49. Griffin, R.C.; Chang, C. Seasonality in Community Water Demand. *West. J. Agric. Econ.* **1991**, *16*, 207–217.

50. Renzetti, S. An Economic Study of Industrial Water Demands in British Columbia, Canada. *Water Resour. Res.* **1988**, *24*, 1569–1573. [CrossRef]

51. Keplinger, K.O.; McCarl, B.A. *The Effects of Recharge, Agricultural Pumping and Municipal Pumping on Springflow and Pumping Lifts Within the Edwards Aquifer: A Comparative Analysis Using Three Approaches*; Texas A&M University: College Station, TX, USA, 1995; Available online: http://agecon2.tamu.edu/people/faculty/mccarl-bruce/papers/584.pdf (accessed on 8 November 2019).

52. Zhao, T.; Dai, A. Uncertainties in Historical Changes and Future Projections of Drought. Part II: Model-simulated Historical and Future Drought Changes. *Clim. Chang.* **2017**, *144*, 535–548. [CrossRef]
53. McCarl, B.A. Cropping Activities in Agricultural Sector Models: A Methodological Proposal. *Am. J. Agric. Econ.* **1982**, *64*, 768–772. [CrossRef]

Article

Climate Change Impacts on Forest Management: A Case of Korean Oak Wilt

Hyunjin An, Sangmin Lee and Sung Ju Cho *

Korea Rural Economic Institute, Naju-si, Jeollanam-do 58321, Korea; hjan713@krei.re.kr (H.A.); smlee@krei.re.kr (S.L.)
* Correspondence: sungjucho@krei.re.kr

Received: 31 October 2019; Accepted: 10 December 2019; Published: 12 December 2019

Abstract: Climate change is expected to affect the occurrence of forest pests. This study depicts a method to measure the impact of damage inflicted by a forest pest like oak wilt as a result of climate change. We determine the damage function considering the factors related to the pest damage and forecast the future damage rate under future climate change. We estimated the damage rate by using the quasi-maximum likelihood estimation (QMLE) and predicted the future damage rate by using representative concentration pathways (RCP) 8.5 data. We assessed the impact of pests on the management income and the rotation age by using a dynamic optimization model. The results show that the damage rate and the affected area from oak wilt would increase under the climate change. In addition, the economic evaluation indicates that altered climate would reduce the management returns and increase uncertainty. However, these outcomes could be alleviated by carrying out the control and prevention measures after the infection occurs.

Keywords: climate change; forest pests; economic impacts; Korean oak wilt; representative concentration pathways

1. Introduction

Climate change, such as an increase in temperature and drought, is expected to directly and indirectly affect the occurrence of pests. Economic losses caused by forest damage in Korea are expected to increase due to the future climate conditions and resulting pests. The insect vector of Korean oak wilt, *Platypus koryoensis*, has emerged since the 2000s. The changed domestic environmental conditions due to climate change made the damage done by the pests severe while the pests had not been a serious concern in the past [1].

Some previous studies examined the relationship between climate change and forest pests in terms of distribution, control, and forest management [2–4]. Other studies used climatic variables, such as seasonal temperature, precipitation, and humidity, as well as other variables such as tree volume, tree health, and pest populations to determine the damage function of forest pests [5–7].

Economic impacts on forest products or services need to consider the effect of time because the growing period is very long for wood. Dynamic optimization can be used to find the optimal rotation age for trees suggested by Faustmann [8] and further suggested by Hartman [9], considering wood as well as non-wood services. Macpherson et al. [10] generalized the Hartman model and showed that when the payout to non-timber value is considered, the rotation age could be shortened or extended relying on the distribution of the pathogenic pests. The result was contrary to the notion that the rotation age is generally believed to decrease when forest pests occur.

We expect to obtain the specific damage rate of forest pests by considering the interaction of host trees, pest occurrence, and climate factors, while the previous studies focused on the potential occurrence and habitat of the pests. Our damage function represents the complex mechanism of pest occurrences by dealing with both the direct and indirect factors that affect pest populations and pests affecting host trees, respectively. In addition, the model includes other factors, such as forest management and human population to assess the impact of human activities. There are few studies examining the economic impact of forest pests in terms of climate change in Korea. While some studies such as An et al. [11] conducted an economic assessment by assuming the damage rate, we use the directly derived damage rate to assess the economic impact. Previous research such as Haight et al. [12] assessed the economic damage from *Ceratocystics fagacearum*, a fungus that causes significant disease of oaks in the central United States using a landscape level model. In their study, the metric of damage is a removal cost. They predict that the discounted damage would be $18-60 million in Anoka County, Minnesota, over the next decade. However, the removal cost is on the lower bounds in total economic loss from the oak wilt because they do not consider the economic losses from reduced services [12]. Our study assesses the economic impact of pests on the management income and rotation age by using a dynamic optimization model. We consider not only the direct impact due to forest pests, such as revenue decrease, but also the revenue change according to managerial factors, such as control and prevention of pests. Lastly, we employ the concept of green payments to cover the indirect value of the environment to deduce a new strategic direction for pest control in forests.

In this paper, we assess the impact of forest pests on the management income and rotation age by using a dynamic optimization model under climate change. In particular, we determine the damage function considering the direct and indirect factors related to the pest damage and forecast the future damage rate under the future climate conditions. Moreover, we evaluate the economic impact of Korean oak wilt: how it changes future return, forest owners, and the optimal rotation age using a dynamic optimization model. We then conduct simulations that deduce the implications for effective pest management in the forest sector. Korean oak wilt is caused by the *Raffaelea* fungus by blocking the nutrient and moisture pathways [3]. *Platypus koryoensis*, a major insect vector of Korean oak wilt, appeared in the 1930s in Korea, but the damage from the Korean oak wilt began to surface in 2004 [13]. The mass damage of oak trees by Korean oak wilt was reported in Gyeonggi Province across a wide distribution of oak trees, and the outbreak had increased rapidly until 2008 [14]. There have been few studies that analyzed the cause of the sudden proliferation of Korean oak wilt. However, the damage in Japanese oak wilt appears mostly in the years with high temperature and low precipitation, so the relationship between the climate and the occurrence of the oak wilt should be paid attention to [14].

We suggest the proliferation factors of Korean oak wilt as the climatic and non-climatic factors by referring to the previous studies. The climatic factors play a direct role in the infection rate by changing the health of trees and the ecology of insect vectors. The non-climatic factors take part in the oak wilt resulting from the host preference of the insect vectors, human activities, and management factors.

The growth of insects, whose habitat was in the southern area of Korea, was promoted by the increasing temperature due to climate change in Korea. The increased temperature may have made the environment advantageous to the insect vector, and thus the damage began to appear [13].

The optimal temperature for growth of the *Raffaelea* fungus inflicting the oak wilt is reported as 25–30 °C [3]. Since the growth of thermophilic species is largely affected by a thermal threshold, their population can significantly decrease when larva is exposed to cold winter temperatures [15]. Experimental report shows that the thermal threshold of an adult flight is 5.8°C [13]. Trees with water stress is usually exposed to attacks by *Platypus koryoensis* because their main target is weak and withered trees [16]. Insect vectors inhabiting weak trees can even attack healthy trees if they are located in nearby areas [17].

The Korea National Park Research Institute (KNPRI) [18] assessed the contribution rate of climatic variables affecting the damage of Korean oak wilt from two national parks in Korea using the maximum entropy model. The results showed that maximum temperature, minimum temperature, and precipitation have high contribution rates, but the average temperature has a low contribution rate. The contribution of maximum temperature is higher than average temperature to the damage of oak wilt According to the research from KNPRI [3], *Platypus koryoensis* preferred to attack the trees with a high DBH. KNPRI collected the sample data from the national park and found that the vectors tend to attack the trees with a 30 cm or larger DBH. The KNPRI survey also confirmed that there is a positive relationship between the damage level and DBH of the trees.

The damage rate of Korean oak wilt also is related to artificial factors, such as the roads, trails, and distance from the village [18]. The experiment data collected from the Bukhansan National Park shows the population of *Platypus koryoensis* was high in the area with a high level of human activity, such as the parking lot and the trail road [18]. However, the author stressed that more detailed research is necessary to determine whether the reason is related to the vector's ecological characteristics, such as a flight habit of *Platypus koryoensis* or the artificial shifting of damaged timber from human activity. In Chiaksan National Park, the damaged trees by Korean oak wilt were concentrated within 20 m of the trail road near Temple Sangwon, where many people frequently visited. They stressed the possibility of *Platypus koryoensis* being infected through the hiking trails in forests. Logging trees also could be a rapid incensement of population of pests [19].

The grass generated by trees attacked by insect vectors releases aggregation pheromone to cause proliferation of damage by group attack. Consequently, failing to manage the damaged trees properly can expand the damage to nearby healthy forests. The insect vectors tend to concentrate on attacking weak trees. Therefore, it is necessary to establish management measures to immediately dispose the withered trees and improve the health of all trees.

2. Materials and Methods

2.1. Pest Damage Function

To measure the damage from pest outbreaks, we used the damage function following the previous study [7]. The damage function of the Korean oak wilt reflecting the factors that affect the insect vectors, pathogens, and host trees can be expressed as follows:

$$D = f_d(Z, P),$$
$$P = f_p(Z, V),$$
$$Z = f_z(W),$$

where
$$D = \text{damage rate, } Z = \text{characteristics of hosts, } P = \text{population of insect vector}$$
$$V, \text{ and } W = \text{exogenous variables.}$$

The pest occurrence (damage) rate D can be represented as the function of the characteristics of the host (Z) and the pest population P. The pest population P is affected by the vector of characteristics of hosts Z and the exogenous variables V such as climatic factors. W is exogenous variables such as precipitation and management factors that affect the tree health. The simultaneous equation can be simplified to the following reduced form as in Cobourn et al. [7]:

$$D = g(W, V).$$

The simplified model is the practical model since exogenous variables V and W are relatively easy to obtain. The possible bias of estimation using W instead of Z may be alleviated if we use a more appropriate explanatory variable W.

The damage rate D can be calculated by following equation:

$$= \frac{Damaged\ area\ (ha)}{Total\ broad\ leave\ forest\ area\ (ha)}, \quad D = [0, 1].$$

The population of insect vectors P is assumed to be a function of the minimum winter temperature of a year ago, relative winter humidity of a year ago, maximum spring temperature, relative spring humidity, maximum summer temperature and its square term, maximum autumn temperature, relative fall humidity, human population, unsalvaged area of damaged tree areas, diameter at breast height (DBH), and dummy variable of the national forest.

We included the climatic variables that directly affect the population of insect vectors. Low minimum winter temperature of last year can cause a decrease in the population of adult beetles since the over-winter larvae tend to be killed under harsh winter conditions. The maximum spring, summer, and autumn temperatures in the current year can affect the flight period of adult beetles. Since the vector beetle is one of the *Platypus* species adapted to warm weather, we chose the maximum temperature, not average temperature, as the variable based on the study by KNPRI [18]. We included the relative humidity in consideration of the fungi that provide food for the larvae. The fungi proliferate more under the hot and humid conditions.

We included the municipality population to investigate the impacts of pest infestation through roads and human movement [20]. The population of the municipality thus can be the instrument variable of infrastructure, such as roads and human movement. The averaged DBH of the trees is included because the insect vectors prefer trees with a large diameter. We also include some management variables, such as the unsalvaged area of trees and national forests. Since the beetle tends to attack the damaged trees, damaged but not salvaged trees may attract more beetles than healthy ones [6]. We also included the national forest (NF_i) as a dummy variable to study the difference of the damage rate according to the management factors. Since national forests are intensively cared for by the government in Korea, rather than private forests, this can be the instrument variable to investigate the impact of human effect on forest insect outbreaks.

The temperature and precipitation are closely related to the health of host trees. The trees are vulnerable to beetle attacks under hot and dry weather since the water stress deteriorates the resistance of host trees [15]. Therefore, we included the precipitation in winter in year $t - 1$ and precipitation in spring, summer, and autumn in year t in the variables. We also included the square term of summer temperature because domestic oak trees are a kind of forest that could be vulnerable under too high temperature.

The period of the data, t, was from 2011 to 2017, and a total of 1610 samples were obtained from 230 municipalities of i, nationwide. The forest-related data, including damaged area by the Korean oak wilt, DBH, and the unsalvaged area of trees in each municipality, were provided by the National Institute of Forest Science. The historical climate data and future climate scenario data of each municipality were obtained from the Climate Change Information Center. Spring is defined as March through May, summer as June through August, autumn as September through November, and winter as December through February in the following year.

2.2. Nonlinear Panel Probit Estimation

The dependent variable in the damage function is the proportional dependent variable that has the value between 0 and 1. Without reflecting the characteristics of the proportional dependent variable, the estimated coefficient can be biased. To reflect such characteristics of the proportional dependent variable, we can express

$$E(y_{it}|x_{it}, c) = \Phi(x_{it}\beta + c_i), \quad t = 1, \ldots, T.$$

Here, the range of the pest damage rate is limited as $0 \leq y_{it} \leq 1$, and the dependent X_{it} is the $1 \times k$ vector. Φ is expressed as the cumulative density function (CDF) of the standard normal distribution and c_i is expressed as the effect between the unobserved cross-section observations. The equation can be expressed by the following equation if we assume exogeneity and the conditional normalized distribution of c_i:

$$E(y_{it}|x_i) \equiv \Phi(\psi + x_{it}\beta + \overline{x_i}\xi).$$

To estimate the equation, we can apply the generalized linear model (GLM) with quasi-likelihood estimation (QMLE) that uses the probit link function [21]. However, inefficiency can be generated since GLM tends to ignore the serial dependence that exists in the joint distribution.

The multivariate weighted nonlinear least square (MWNLS) is known to be ideal to estimate the panel data that has the serial dependence and heteroscedasticity. However, it is very difficult to estimate the parametric model to $Var(y_i|x_i)$ [21].

To supplement the weakness, Papke and Wooldridge [21] suggested using the quasi-maximum likelihood estimation (QMLE) instead of finding the parametric model. When correctly specified, MWNLS and QMLE become the asymptotically equivalent estimation. This study that has the panel data and proportional dependent variable utilized the QMLE that applied the probit link function and robust standard errors.

The below equation shows the conditional average of the pest damage rate of this study that has N municipalities ($i = 1, \ldots, N$) and T years ($t = 1, \ldots, T$).

$$E(y_{it}|x_i) \equiv \Phi(\psi + x_{it}\beta + \overline{x_i}\xi).$$

Here, the variables y_{it}, x_{it}, and $\overline{x_i}$ refer to the pest damage rate (dependent variable), climatic factors and non-climatic factors (explanatory variable), and the average of the panels of the explanatory variables, respectively. Since it is difficult to analyze the estimation coefficients of the nonlinear model estimated with QMLE, we should deduce the average marginal effect (AME) which means the change of the dependent variable affected by the change of a unit of the explanatory variable [21]. We can observe the effect of the change of a unit of the dependent variable on the pest damage rate using the AME:

$$AME_k = N^{-1} \sum_{i=1}^{N} \widehat{\beta_k}\phi(\hat{\psi} + x_i\hat{\beta} + \overline{x_i}\hat{\xi}).$$

Here, ϕ means the probability density function (PDF) for the standard normal distribution.

2.3. Projections of Korean Oak Wilt Climate Change

We used the representative concentration pathways (RCP) 8.5 data provided by the Korea Meteorological Administration (KMA) to forecast the future damage rate according to climate change in the Korean Peninsula. The forecast measured the dependent variable (damage rate) by applying the future weather data to the estimated coefficient. We assumed the non-climatic variables to be the same as in 2018 and created the future data through the assumption. The damage rates of the Korean oak wilt in each municipality in South Korea from 2018 to 2020 were calculated in the process.

We assumed that the population and the area of infected trees without control were the same as 2018. Although it may not be consistent with the declining population trend in Korea, this study focuses on the correlation between climate and oak wilt rather than the artificial factors such as the population. For the DBH, we applied the average DBH change rate with reference to the "Timber Biomass and Harvesting Table" published by NIFS [22]. In other words, we added the average DBH change rate every 10 years to the average DBH observed in 2018 to obtain the DBH change until 2100.

2.4. Economic Evaluation of Korean Oak Wilt

The following equation developed by Macpherson shows the objective function to assess the forest value that includes the timber and the non-timber incomes [10]:

$$\max_{t} PV(t)Le^{-rt} - CL + \int_{0}^{t} G(L)e^{-rs}ds + \int_{t}^{\infty} aLe^{-rs}ds,$$

where the L is forest area, and $G(L)$ implies the green payment that is assumed to be a function of the forest area. Then the optimal condition can be expressed as the following equation:

$$\frac{V'(t)}{V(t)} - r = \frac{1}{L}\frac{aL - G(L)}{PV(t)}.$$

We assume areas producing timber value affected by oak wilt outbreaks is expressed as L^i_{TB}, and the area generating non-timber value affected by oak wilt outbreaks is expressed as L^i_{NTB}. L^i_{NTB} can be divided into n small sections. Then the total area of producing timber value would be expressed as the following equation:

$$L^i_{NTB} = \sum_{i=1}^{n} \sigma_i x_i, \quad 0 \le \sigma_i \le 1.$$

Therefore, the objective function including the timber and non-timber values can be expressed as follows:

$$\max_t PV(t) L^i_{TB}(t) e^{-rt} - CL + \int_0^t G\big(L^i_{NTB}(s)\big) e^{-rs} ds + \int_t^\infty aL e^{-rs} ds.$$

Since the green payment $G\big(L^i_{NTB}(t)\big)$ is granted to the area that creates the non-timber value, the total green payment can be calculated by the following equation where g is the green payment per unit area. This represents the non-timber values from the forests.

$$G\big(L^i_{NTB}(t)\big) = g \times L^i_{NTB}(t).$$

Taking first differentiation to the above equation with respect to time (t), the condition for the optimal forest rotation age can be expressed as follows:

$$\frac{V'(t)}{V(t)} - r = \frac{1}{L^i_{TB}(t)} \left(\left| \frac{dL^i_{TB}(t)}{dt} \right| + \frac{1}{PV(t)} \left(aL - e^{rt} \frac{d}{dt} \left(\int_0^t G\big(L^i_{NTB}(s)\big) e^{-rs} ds \right) \right) \right).$$

Finally, if we include the control and prevention of oak wilt, the objective function can be expressed as the following. The purpose of the objective function is to find the optimal rotation age (t), which generates the best present value for the cost of the control and prevention to the oak wilt outbreaks. We assume the non-timber value is provided annually, and the timber values are generated during harvest.

$$\max_t PV(t) L^c_{TB}(t) e^{-rt} - CL + \int_0^t \big[G\big(L^c_{NTB}(s)\big) - D(I(s)) \big] e^{-rs} ds + \int_t^\infty aL e^{-rs} ds.$$

The optimal condition can be expressed as follows:

$$\frac{V'(t)}{V(t)} - r = \frac{1}{L^c_{TB}(t)} \left(\left| \frac{dL^c_{TB}(t)}{dt} \right| + \frac{1}{PV(t)} \left(aL - e^{rt} \frac{d}{dt} \left(\int_0^t \big[G\big(L^c_{NTB}(s)\big) - D(I(s)) \big] e^{-rs} ds \right) \right) \right).$$

When no action has been carried out in the infected area, the forest area can be separated into two classes: the susceptible region (S(t)) and the infected region (I(t)). That is, the total area L, the sum of S(t) and I(t). If the forest areas affecting timber and non-timber return (L^i_{TB} and L^i_{NTB}) are denoted as follows:

$$L^i_{TB}(t) = S(t) + \rho(L - S(t)),$$

where (L-S(t)) in the right-hand side implies the infected area (I(t)). When the control and preventive measures are applied, timber areas can be separated into three classes: the susceptible area (S(t)), the controlled area (T(t)), and the infected area (I(t)). Thus, the area can be expressed as the following if control and prevention are included:

$$L^c_{TB}(t) = S(t) + (\alpha + \rho) \frac{L - S(t)}{1 + \alpha},$$

$$L^c_{NTB}(t) = S(t) + (\alpha + \sigma) \frac{L - S(t)}{1 + \alpha},$$

where the controlled area (T(t)) is assumed to be free from pest infestation or timber production and to be linearly proportional to the infected area ($T(t) = \alpha I(t)$) with the control rate α.

The data needed for a numerical assessment using the above model involve timber production function, changes in the pest infestation area over time, annual land area, damage rates, costs for control and prevention, timber prices, and costs for logging and afforestation. Most of the data are publicly accessible, but the pest infection area over time can be gained by using the SI model (susceptible–infected model). The SI model for the no-action model can be denoted as follows:

$$\frac{dS}{dt} = -\beta S(t)(I(t) + p),$$
$$\frac{dI}{dt} = \beta(S(t)(I(t) + p),$$

with p referring to the initial infected area and β referring to the secondary infection rate within the forest. The total forest area (L) is expressed as the sum of S(t) and I(t) if no control measures to the oak wilt have been carried out, and the change of S(t) with respect to time is described as follows:

$$\frac{dS}{dt} = -\beta S(t)(L - S(t) + p).$$

Applying the variable separation method for the solution of the above differential equation, S(t) can be described as follows:

$$S(t) = \frac{L + p}{(p/L)e^{(L+p)\beta t} + 1}.$$

In models for control and prevention (L is separated into S(t), T(t), and I(t)), changes in S(t) with respect to time are expressed as follows:

$$\frac{dS}{dt} = -\beta S(t)\left(\frac{L - S(t)}{1 + \alpha} + p\right).$$

Similarly, S(t) yields the following equation by using the variable separation:

$$S(t) = \frac{L + p(1 + \alpha)}{\frac{p(1+\alpha)}{L}\exp\left((L + p(1 + \alpha))\frac{\beta t}{1+\alpha}\right) + 1}.$$

3. Results

3.1. Estimated Damage Function

Table 1 shows the coefficients and average marginal effects of the factors on the damage rate of Korean oak wilt. The results show that the climatic factors are related to the proliferation of Korean oak wilt, including the minimum last winter temperature and precipitation, maximum spring temperature, maximum summer temperature and precipitation, and relative humidity in autumn. In general, increasing temperature is expected to extend the outbreak rate of the oak wilt. Because the relationship between the damage rate and the average marginal effect of the minimum winter temperature and maximum spring temperature is linear with a positive sign, the damage rate from the oak wilt is likely to extend by increasing the minimum winter temperature and maximum spring temperature on average. Although the marginal effects of the maximum summer temperature show a negative sign on average, its quadratic form leads to the marginal effect gradually decreasing at 27 °C or higher and approaching 0 at 35 °C or higher as in Figure 1.

Table 1. Estimation results of the Korean oak wilt damage function.

	Coef.			AME		
Infected area without control (ha)	0.0118	(0.0038)	***	0.000031	(0.000011)	***
Diameter (cm)	0.0160	(0.0047)	***	0.000042	(0.000014)	***
Population (million)	2.6337	(0.7739)	***	0.006906	(0.002222)	***
Minimum last winter temperature (°C)	0.0814	(0.0335)	**	0.000213	(0.000089)	**
Precipitation in last winter (mm)	−0.0219	(0.0042)	***	−0.000058	(0.000014)	***
Relative humidity in last winter (%)	0.0076	(0.0087)		0.000020	(0.000023)	
Maximum spring temperature (°C)	0.1269	(0.0534)	**	0.000333	(0.000130)	**
Relative humidity in spring (%)	0.0003	(0.0025)		0.000001	(0.000007)	
Precipitation in spring (mm)	0.0003	(0.0257)		0.000001	(0.000067)	
Maximum summer temperature (°C)	0.9137	(0.7711)		−0.000382	(0.000207)	*
Max. summer temperature squared	−0.0181	(0.0131)				
Precipitation in summer (mm)	−0.0014	(0.0008)	*	−0.000004	(0.000002)	*
Maximum autumn temperature (°C)	0.0352	(0.0562)		0.000092	(0.000151)	
Relative humidity in autumn (%)	0.0098	(0.0228)		0.000026	(0.000061)	
Precipitation in autumn (mm)	−0.0066	(0.0018)	***	−0.000017	(0.000006)	***
National forest (1 = Yes; 0 = No)	0.0389	(0.0901)		0.000105	(0.000251)	
Constant	−26.8529	(11.0424)	**			
Number of observations	2512					
Number of clusters	314					
Pseudo R^2	0.2019					
Log-likelihood	−14.5147					

Note: Clustered robust standard errors are in parentheses. *, **, *** indicate statistical significance at the levels of 10%, 5%, and 1%, respectively. Coef. and AME indicate coefficients and average marginal effects, respectively.

Precipitation in general shows a negative relationship with the damage rate. Pertaining to the seasonal factors, the average precipitation in winter, summer, and autumn are statistically significant at the 1% and 10% levels, respectively. The decrease in precipitation affects moisture stress and reduces the resistance of the host trees, and thus it is prone to extend the damage due to the rapid proliferation of the oak wilt. Decreasing precipitation in the winter, autumn, and summer is likely to increase the damage rate from the oak wilt.

Previous studies have shown that increasing temperature leads to the increase in the active period of the insect vector because of the decreasing death rate of larvae and early eclosion of adults. These studies show that the proliferation of oak wilt is directly affected by the increased activity and population of the insect vector. Our estimation results are consistent with these studies in that the increasing temperature is likely to extend the damage rate and the high summer temperature could lead to a reduced infection rate thanks to slowed spawning or the migration of the insect vector.

Figure 1 shows the conditional mean of the damage rate per ha and the 95% confidence intervals pertaining to the level of some climatic variables. Increasing the confidence intervals indicates that the uncertainty is likely to increase as the maximum spring temperature increases and the precipitation in summer and autumn decreases. However, the uncertainty due to the large interval is not likely to be significant under the RCP 8.5 scenario predicting the 4 °C rise in temperature by 2100.

The non-climatic factors such as the infected area without control, diameter at breast height (DBH), and population are also significantly associated with the damage rate. The estimation results demonstrate the positive relationship between the damage rate and the infected area without control, DBH, and population. As the infected area without control increases, the damage to nearby healthy forests is proliferated because of the pheromone emitted by the insect vector of the infected trees that attracts the other insect vectors nearby. Gan [6] also showed the positive relationship between the damage rate and the infected trees without control. Thus, dealing with the infected trees properly is likely to alleviate the proliferation of the damage rate while neglecting them can worsen the damage rate.

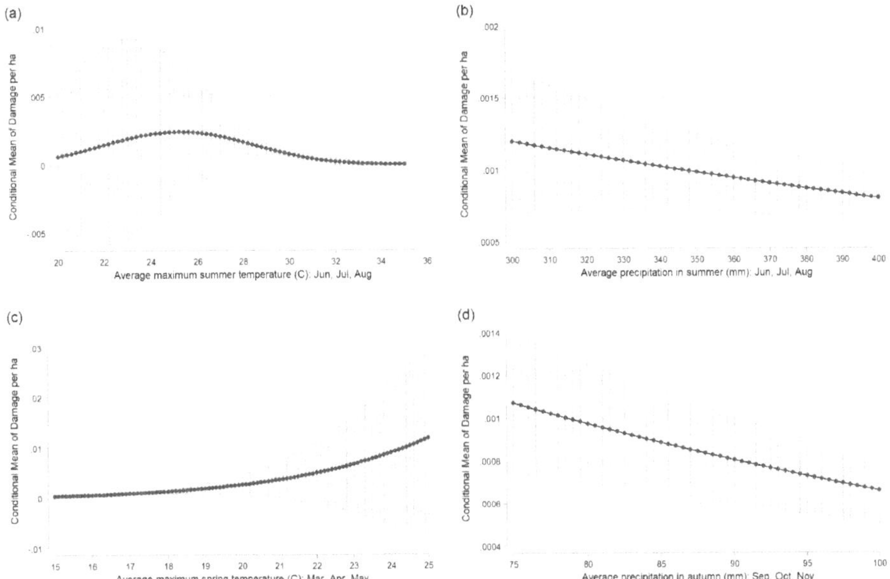

Figure 1. Average marginal effect by major climatic variables. (**a**) Average maximum summer temperature; (**b**) average precipitation in summer; (**c**) average maximum spring temperature; (**d**) average precipitation in autumn.

The estimation result that the damage rate increases as DBH increases is consistent with the result of existing studies that the damage mostly occurs in large trees. In other words, the large trees are more likely to be attacked by *Platypus koryoensis*.

We employed the population as the proxy variable showing human activities and infrastructure such as roads. The results show that the population is positively associated with the damage rate. Previous studies have shown that the population of insect vectors in regions with roads and trails, with a high floating population, was larger than the areas difficult to access, such as forests.

The pest control in national forests is known to be better than that in the municipality or private sector, but the marginal effect of the national forest on the damage rate is not statistically significant. This may be because the unit of the panel is the municipalities and the large difference in the number of samples among the forests by owner type.

3.2. Projection Results

Figure 2 demonstrates the forecast of the damage rate from the Korean oak wilt under the RCP 8.5 scenario. For the period of 2011–2017, the Korean oak wilt occurs in Seoul and Gyeonggi Province that are the most populated areas in Korea. The affected areas are likely to be expanded to not only further north but also to the east and west coastal areas. From the 2050s, the affected areas are expected to gradually expand to South Gyeongsang Province, the coastal areas of Chungcheong regions, and some coastal areas of Gangwon Province are likely to be affected by the oak wilt in the 2090s. Although insect outbreak may be affected by biological factors, such as the natural enemies of pathogens and resources for insects, our model mostly focuses on impacts of climate and human intervention on Korean oak wilt outbreaks. Future research considering complicated biological characteristics may be conducted to improve the estimates of the projected damage rates.

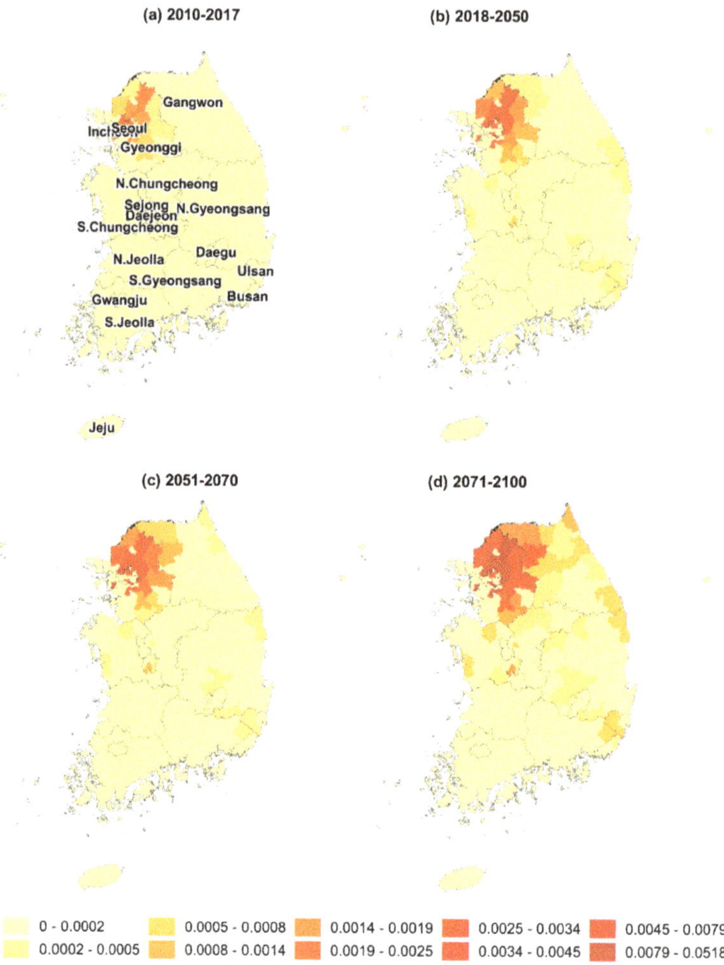

Figure 2. Forecast of the damage rate from Korean oak wilt.

3.3. Economic Evaluation of Korean Oak Wilt

To analyze the economic evaluation of Korean oak wilt, we utilized the forest harvest table of NIFS [22] and chose *Quercus acutissima*, *Quercus variabilis*, and *Quercus mongolica*. Table 2 illustrates the parameters for setting the baseline with p referring to the area initially infected and β referring to the secondary infection rate. The p per area (ha) was estimated to be 0.00087 by using the 2010 data of the infected areas. We calculated the β value using the equation $dI/dt = \beta S(t)(I(t) + p)$ with the infected area (I(t)) and susceptible area (S(t)) data of the region in the 2012–2017 period, which results in 0.0017. We assumed that the infected area with possible use of timber (ρ) and the infected area with possible use of non-timber (σ) are 0.5, which indicates that the infected trees are assumed to lose half of their timber and non-timber value. We used age-specific volumes by using the surveyed data of tree age from the "Timber Biomass and Harvesting Table" published by NIFS [22] instead of the volume production function. Data for timber price (KRW 1000/m3), planting cost (KRW 1000/m3), and afforestation cost (KRW 1000/m3) were obtained from Min et al. [23]. We assumed that the oak wilt appears in 10 year or older trees due to the preference of the insect vector of oak wilt for large trees and that the trunk injection is targeted to trees over 10 cm in DBH, which are in general 15 to 20 years old.

Table 2. Parameters for the forest rotation age analysis.

	Unit	*Quercus acutissima*	*Quercus variabilis*	*Quercus mongolica*
Market price of timber	KRW 1000/m3	83.5	83.5	83.5
Cost of planting	KRW 1000/ha	8339	8339	8339
Cost of logging	KRW 1000/ha	16,109	16,109	16,109
Green payment	KRW 1000/ha	100	100	100
Discount rate	%	3	3	3
p		$0.00087 \times L$	$0.00087 \times L$	$0.00087 \times L$
β		0.0017	0.0017	0.0017
ρ		0.5	0.5	0.5
σ		0.5	0.5	0.5
Cost of control	KRW 1000/ha	980	980	980
Cost of handling withered trees	KRW 1000/ha	2200	2200	2200
Area (L)	ha	100	100	100

Table 3 shows the forest rotation ages decrease when the trees are infected compared to the no infection case (44–70 years). Furthermore, the rotation age is likely to be shortened when the pest control and prevention is not carried out (33–44 years) than when the measure is carried out (41–59 years).

Under the given condition, it is difficult to expect positive returns through forest management because the present values of objective function are negative in every case. However, the objective function value is highest in the case of no infection, and the value is much higher in the case of the control measures than in the case of no control measures when the oak wilt occurs. It indicates that the cost of pest control and preventive measures are less than the cost of losing timber and non-timber values.

Table 3. Change in forest rotation age due to the infection of Korean oak wilt (units: years, KRW).

	No Infestation		Infestation, No Control		Infestation, Control	
	Rotation Age	Present Value	Rotation Age	Present Value	Rotation Age	Present Value
Quercus variabilis	44	−352,830	33	−462,166	41	−378,795
Quercus acutissima	70	−383,332	44	−658,351	59	−478,286
Quercus mongolica	67	−382,364	36	−575,852	54	−436,174

3.4. Simulation

We conducted a simulation on *Quercus variabilis* according to the parameter values and examined the impact of parameter changes from the baseline on the rotation age and the objective function value. We then deduced policy implications based on the simulation results.

3.4.1. Changes in the Market Price of Timber

The simulation results in Figure 3 show that as the timber market price increased (KRW 80,000–200,000/m^3). The forest rotation age of *Quercus variabilis* is expected to reduce from 45 to 31 years when no infection occurs and from 33 to 28 years when no control is carried out after the infection occurs. The changes of the forest rotation age when the control measures are carried out are similar to the changes when no infection occurs, and the rotation age in the two cases are similar at the price of timber of KRW 180,000/m^3 or higher.

Figure 3 also illustrates that forest management return reduces if pest infection occurs, but the value increases if the control measures are carried out and becomes close to the value of the case of no infection. The objective function value turns to a positive value when the price of timber is around KRW 160,000/m^3 in all cases.

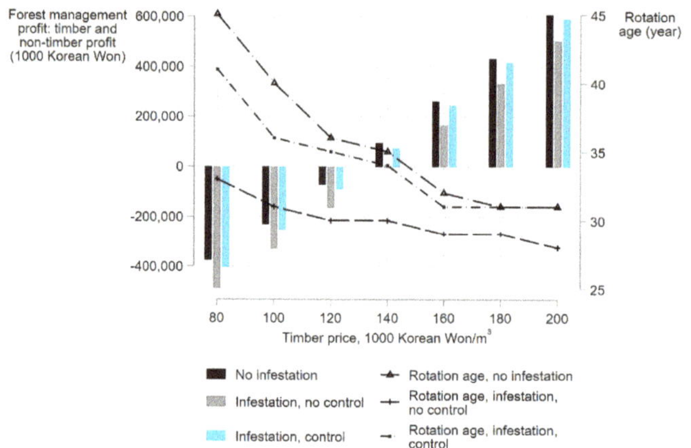

Figure 3. Change in management returns of *Quercus variabilis* under the change of timber price.

3.4.2. Changes in Green Payments

Assuming that the green payments are paid to the forest owners with the amount of KRW 100,000/ha, the simulation results in Figure 4 show the forest rotation age of 44, 33 (infection with no control), and 41 years (infection with control). Figure 4 illustrates that the forest rotation age and the forest management returns are expected to gradually increase as the amount of green payments increases. The forest owners are likely to have the incentive to preserve the trees and forests as the green payments become higher. The results also indicate the difference of rotation age between the cases of control and no control after the infection, in which the increasing green payments are far more significant in the increases in the rotation age and the management returns. In the case of no infection and infection with control measures, the forest management returns turn to positive value with the green payments of about KRW 300,000/ha.

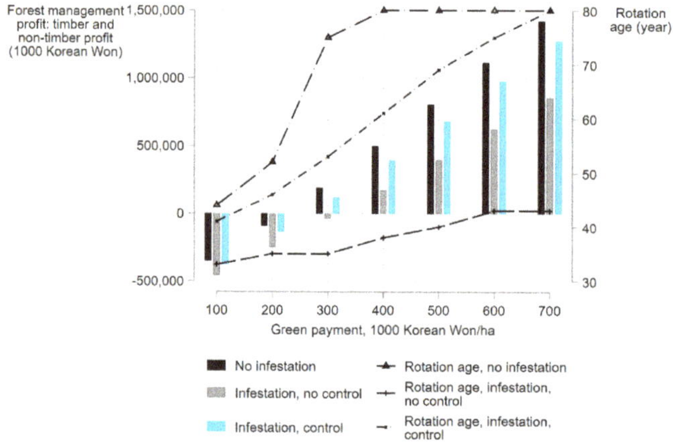

Figure 4. Change in management returns of *Quercus acutissima* according to the green payment.

3.4.3. Changes in Climate

Considering the estimation and projection results, the increasing damage rate due to climate change is expected to play a negative role in the economic returns in the forest as well. For the simulation, we estimated the economic impact of the adjustment of the β value corresponding to the outbreak rate under altering climate. We derived the forest rotation age and the forest management returns from timber and non-timber that satisfied the optimization condition by using the estimated β value as described above. Table 4 demonstrates the average of β of the Korean oak wilt for the 30-year periods.

Table 4. Change of Korean oak wilt breakout rate (β).

Period	2011–2040	2041–2070	2071–100
β value	0.0016	0.0023	0.0044

In Figure 5, the forest rotation age reduces as β increases. Although the change of the returns is relatively small, the deviation increases significantly when there is no control measure (KRW 300 million) than when there are control measures (KRW 90 million) after the infection. In this case, income of forest owners can be stabilized with the control and prevention measures under the changes in climate.

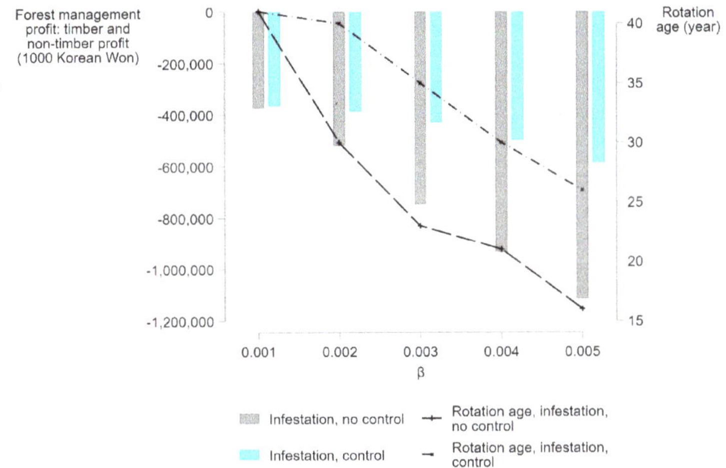

Figure 5. Change of management returns under the change of Korean oak wilt outbreak rate (β).

3.4.4. Change in Utilization Rate of Infected Trees

Figure 6 shows the impacts of the utilization rate (ρ and σ) affecting the production of timber and non-timber on the rotation age and the forest management returns. We assumed that the ρ and σ values are the same as in the baseline. When the impact of the Korean oak wilt is high with small ρ and σ, the decrease of the returns with no control is higher than that with control measures. However, the return gap between the cases narrows as the utilization rate increases because a part of the damage from the Korean oak wilt can be offset by the high utilization rate.

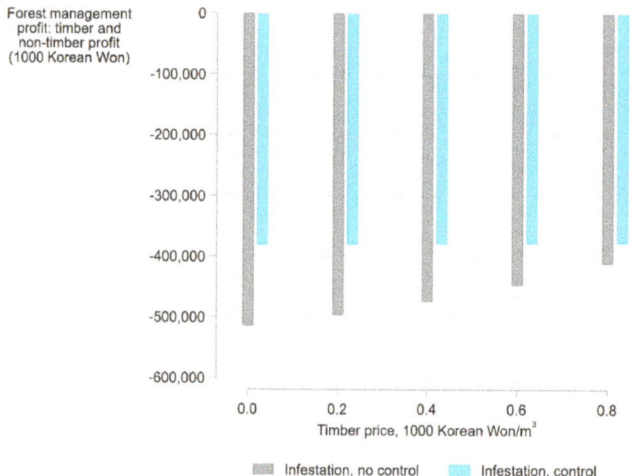

Figure 6. Change of management returns of *Quercus variabilis* under the change of utilization rate (ρ, σ) of infected trees.

4. Discussion

4.1. Estimation and Projections

The estimation result of the damage function of Korean oak wilt indicates that the damage rate is positively correlated with the minimum winter temperature, maximum spring temperature, and the linear term of the maximum summer temperature. In contrast, the damage rate is negatively correlated with the precipitation in winter, summer, and autumn, as well as the square term of the maximum summer temperature. It also indicates that the damage rate is likely to increase as temperature increases, but in exceedingly high summer temperature it is likely to begin to decrease

The non-climatic factors including the infected area without control, DBH, and population play a significant role in the damage rate from Korean oak wilt. The damage rate from Korean oak wilt extends as the area of infected trees without control increases, and when the diameter and population increase. The estimation results imply that the damage rate is more affected by indirect causes, such as the condition of host trees and managerial factors, than the insect vector population. Gan [6] also shows that management factors have a significant impact on forest pests. His research investigates the relationship between various factors, such as climate and management, as well as southern pine beetle (SPB) infestation using the panel data mode. He found a positive relationship between SPB and unsalvaged timber volume. Thus, improving the tree health through preventive measures can help prevent the Korean oak wilt

Projection results of the future damage rate indicate that the affected regions are likely to expand further north and to coastal areas in the east and the west while the current pest occurrence is concentrated on the capital and nearby regions. As in the case of Japan in which the Japanese oak wilt mainly expanded to the coastal areas, the damage in the coastal regions in Korea should be paid close attention to as well.

4.2. Economic Analysis

In the case of sustainable forests, there should be positive returns on forest management, which may be done with significant increases in timber price or payout for various values of forests. Under the current condition, however, it is difficult to expect a positive income even by considering both timber and non-timber values of forests. Our results show that it is better not to use the forest in terms of profit, and the profits are declined due to the Korean oak wilt infection. The economic evaluation results show the importance of pest control and prevention measures because the economic returns and rotation age deteriorates when there is not a control measure after the infection, even with increases in timber price, green payments, and utilization rate.

Under climate change, it is expected that the probability of Korean oak wilt infection would intensify the decreases in the management returns and the increases in the uncertainty. No control measures also exacerbate the income stability when the Korean oak wilt occurs. The forest rotation age shortens when the damage rate increases due to climate change. Previous research also shows rotation age is reduced when it maximizes the net present value of forest and trees are damaged by pests. Macpherson et al. [24] show that the rotation age is shortened when timber from infected trees has no value (only timber of undamaged trees would be sold) and the faster the infestation spreads the shorter the optimal rotation age. Increasing the risk of a catastrophic loss such as forest pest and fire shortens the optimal rotation age [25]. Reed [26], adapting the infinite rotation Faustmann formula to the arrival of fire, found that the risk of catastrophic event shortens the optimal rotation age due to increasing the effective discount rate. Therefore, the forest owners notice a higher opportunity cost of not harvesting.

Thus, active control and prevention measures with additional support for income stabilization would help prepare for increasing pest occurrence and keep forests sustainable. Our simulation results indicate that the impacts of economic support, such as green payments and increasing the timber price, would decline further in the future if the forest management returns decrease due to Korean oak wilt. Thus, the policy to stabilize and/or increase income of forest owners would be more effective when it is applied before the effects of climate change take place.

5. Conclusions

Climate change in the long run establishes the environmental circumstance favorable to pests that competently adapt to changing environments. As the favorable region for inhabitation expands, the damage is also expected to expand in new areas not previously affected by Korean oak wilt. New damages are expected on the west coast and the southern region of the east coast after the 2050s. Since Japanese oak wilt extended around the west coast in Japan in the 1990s, it is advisable to monitor the Korean oak wilt occurrence especially in the coastal areas in Korea. The predicted increase in winter temperature is expected to cause increasing damage in the cold and mountainous regions which were not affected by the damage before. The forest management returns would deteriorate if the infection of Korean oak wilt intensifies compared to when there was no oak wilt. Furthermore, this study demonstrates that the management returns worsen faster, especially in the case of no pest control after the infection. Thus, control and preventive measures are necessary to protect the income of forest owners. However, most of the Korean forest owners do not have the incentive to control forest pests because of the low economic efficiency of domestic forests. Therefore, it is necessary to provide an incentive for forest owners, such as the green payments, to actively participate in pest control and expand their role as the actual control subject. Since the estimation results show that moisture stress is one of the key factors in deteriorating tree health, it is necessary to pay attention to supplying water during the dry weather.

Author Contributions: Data curation, S.J.C.; methodology, H.A.; project administration, H.A.; software, S.J.C.; supervision, S.L.; writing—original draft, H.A. and S.L.; writing—review and editing, S.J.C.

Funding: This research was funded by Korea Rural Economic Institute (No. R840).

Acknowledgments: This research is a modified version of the report entitled "The Effects of Climate Change on Forest Insect Disturbance in South Korea—Challenges and Prospects" that was published by Korea Rural Economic Institute in 2018.

References

1. Lee, S.-W.; Um, T.-W. Study on the Current State and Control of Oak Wilt Infected According to Climate Change. *Proc. Korean Soc. Environ. Ecol. Con.* **2014**, *2014*, 23–24.
2. Sturrock, R.N.; Frankel, S.J.; Brown, A.V.; Hennon, P.E.; Kliejunas, J.T.; Lewis, K.J.; Worrall, J.J.; Woods, A.J. Climate change and forest diseases. *Plant Pathol.* **2011**, *60*, 133–149. [CrossRef]
3. The Korea National Park Research Institute. *Oak Wilt Disease Outbreak and Prediction in Bukhansan National Park*; Korea National Park Service: Wonju-si, Korea, 2010.
4. Lim, J.H.; Kwon, T.S.; Chun, J.W.; Park, G.E.; Choi, W.I.; Choi, H.T.; Kim, S.; Chul, C.; Shin Kyu, K.; Dong-wook, K.; et al. *Forest Ecosystem Impact Assessment and Adaptation to Climate Change*; National Institute of Forest Science: Seoul, Korea, 2015.
5. Kroll, J.C.; Reeves, H.C. A Simple Model for Predicting Annual Numbers of Southern Pine Beetle Infestations in East Texas. *South. J. Appl. For.* **1978**, *2*, 62–64. [CrossRef]
6. Gan, J. Risk and Damage of Southern Pine Beetle Outbreaks under Global Climate Change. *For. Ecol. Manag.* **2004**, *191*, 61–71. [CrossRef]
7. Cobourn, K.M.; Burrack, H.J.; Goodhue, R.E.; Williams, J.C.; Zalom, F.G. Implications of simultaneity in a physical damage function. *J. Environ. Econ. Manag.* **2011**, *62*, 278–289. [CrossRef]
8. Faustmann, M. Calculation of the Value Which Forest Land and Immature Stands Possess for Forestry. *Allg. Forst Jagdztg.* **1849**, *15*, 441–455.
9. Hartman, R. The Harvesting Decision Whena Standing Forest Has Value. *Econ. Inq.* **1976**, *14*, 52–58. [CrossRef]
10. Macpherson, M.F.; Kleczkowski, A.; Healey, J.R.; Hanley, N. Payment for multiple forest benefits alters the effect of tree disease on optimal forest rotation length. *Ecol. Econ.* **2017**, *134*, 82–94. [CrossRef] [PubMed]
11. HyunJin, A.; Sung Ju, C.; Sera, O.; Jae Min, J. Economic Impacts of Invasive Pests under Climate Change: A Case of Lycorma delicatula. *J. Korea Acad.-Ind. Coop. Soc.* **2018**, *19*, 415–422.
12. Haight, R.G.; Homans, F.R.; Horie, T.; Mehta, S.V.; Smith, D.J.; Venette, R.C. Assessing the cost of an invasive forest pathogen: A case study with oak wilt. *Environ. Manag.* **2011**, *47*, 506–517. [CrossRef] [PubMed]
13. Nam, Y.; Koh, S.-H.; Won, D.-S.; Kim, J.-K.; Choi, W.I. An empirical predictive model for the flight period of Platypus koryoensis (Coleoptera: Platypodinae). *Appl. Entomol. Zool.* **2013**, *48*, 515–524. [CrossRef]
14. Korea Forest Research Institute. *Report of Monitoring for Forest Insect Pests and Diseases in Korea*; Korea Forest Research Institute: Seoul, Korea, 2011.
15. Bentz, B.J.; Régnière, J.; Fettig, C.J.; Hansen, E.M.; Hayes, J.L.; Hicke, J.A.; Kelsey, R.G.; Negrón, J.F.; Seybold, S.J. Climate Change and Bark Beetles of the Western United States and Canada: Direct and Indirect Effects. *BioScience* **2010**, *60*, 602–613. [CrossRef]
16. Bentz, B.J.; Logan, J.A.; Amman, G.D. Temperature-Dependent Development of The Mountain Pine Beetle (Coleoptera: Scolytidae) and Simulation of Its Phenology. *Can. Entomol.* **1991**, *123*, 1083–1094. [CrossRef]
17. Gaylord, M. *Climate Change Impacts on Bark Beetle Outbreaks and the Impact of Outbreaks on Subsequent Fires*; Ecological Restoration Institute and Southwest Fire Science Consortium, Northern Arizona University: Flagstaff, AZ, USA, 2014.
18. The Korea National Park Research Institute. *Spread Prediction Analysis of Oak Wilt Disease in Gyeryongsan National Park*; Korea National Park Service: Wonju-si, Korea, 2012.
19. Ciesla, W. *Forest Entomology: A Global Perspective*; John Wiley & Sons: Hoboken, NJ, USA, 2011; ISBN 978-1-4443-9788-8.
20. Roques, A.; Zhao, L.; Sun, J.; Robinet, C. Pine wood nematode, pine wilt disease, vector beetle and pine tree: How a multiplayer system could reply to climate change. In *Climate Change and Insect Pests*; CABI Climate Change Series; CABI: New York, NY, USA, 2015; pp. 220–234. ISBN 978 1 84593635 8.
21. Papke, L.E.; Wooldridge, J.M. Panel data methods for fractional response variables with an application to test pass rates. *J. Econom.* **2008**, *145*, 121–133. [CrossRef]
22. *National Institute of Forest Science Timber Resources, Biomass and Standing Forest (임목재적·바이오매스 및 임분 수확표)*; Korea Forest Service: Daejeon, Korea, 2012; ISBN 978-89-8176-875-1.
23. Min, K.; Seok, H.; Choi, J. *Policy Tasks to Improve the Profitability of Forest Management in Korea*; Korea Rural Economic Institute: Naju-si, Korea, 2018.

24. Macpherson, M.F.; Kleczkowski, A.; Healey, J.R.; Hanley, N. The Effects of Disease on Optimal Forest Rotation: A Generalisable Analytical Framework. *Env. Resour. Econ.* **2018**, *70*, 565–588. [CrossRef] [PubMed]
25. Amacher, G.; Ollikainen, M.; Koskela, E.A. *Economics of Forest Resources*, 1st ed.; The MIT Press: Cambridge, MA, USA, 2009; ISBN 978-0-262-01248-5.
26. Reed, W.J. The effects of the risk of fire on the optimal rotation of a forest. *J. Environ. Econ. Manag.* **1984**, *11*, 180–190. [CrossRef]

Article

Achieving Food Security in a Climate Change Environment: Considerations for Environmental Kuznets Curve Use in the South African Agricultural Sector

Saul Ngarava [1,*]**, Leocadia Zhou** [1]**, James Ayuk** [1] **and Simbarashe Tatsvarei** [2]

[1] Risk and Vulnerability Science Centre, University of Fort Hare, Alice 5700, South Africa
[2] Department of Agri-business and Entrepreneurship, Marondera University of Agricultural Sciences and Technology, Marondera 00263, Zimbabwe
* Correspondence: ngaravasaul@gmail.com; Tel.: +27-73-203-7094

Received: 22 July 2019; Accepted: 27 August 2019; Published: 3 September 2019

Abstract: This study relates agricultural income and agricultural carbon dioxide (CO_2) emissions in the context of environmental Kuznets curves for South Africa. We posit likely relationships between UN Sustainable Development Goals (SDG) 1, 2 and 13, relating food production to climate change action. CO_2 emissions, income, coal energy consumption and electricity energy consumption time series data from 1990 to 2012 within the South African agricultural sector were used. The autoregressive distributive lag bounds-test and the error correction model were used to analyse the data. The results show long-run relationships. However, agricultural income was only significant in the linear and squared models. Changes in agricultural CO_2 emissions from the short run towards the long run are estimated at 71.9%, 124.7% and 125.3% every year by the linear, squared and cubic models, respectively. Exponentially increasing agricultural income did not result in a decrease in agricultural CO_2 emissions, which is at odds with the Kuznets hypothesis. The study concludes that it will be difficult for South Africa to simultaneously achieve SDGs 1, 2 and 13, especially given that agriculture is reliant upon livestock production, the largest CO_2 emitter in the sector. The sector needs to shift to renewable energy consumption with fewer CO_2 emissions.

Keywords: agriculture; carbon dioxide; environmental Kuznets curves; South Africa; sustainable development goals

---

1. Introduction

Climate change has become a topical issue, as witnessed by its inclusion in the United Nations' 2030 Agenda Sustainable Development Goals (SDGs), which are a global plan of action for people, the planet and prosperity, seeking to eradicate poverty. Sustainable Development Goal (SDG) 13 pertains to combating climate change and its impacts [1,2]. Climate change is associated with changes in ambient CO_2 concentrations [3–5]. In achieving SDG 13 and reducing the impact of climate change, transformative policies and actions are required for the reduction of CO_2 emissions. These transformative actions, however, come with their own downsides. These include trade-offs with productive capacities, especially for developing countries [1,6]. Highlighting such downsides will likely lead to short-term and long-term impacts. Agriculture is one of the primary sectors that is affected by climate change, with its impact being both spatial and temporal in scale [1]. The sector is both a perpetrator and a victim of climate change. Agriculture is the primary activity for 2.5 billion people worldwide [3]. According to van Noordwijketal [7], agriculture is essential for achieving the SDGs through the interaction of three broad categories of the SDGs, namely its redistributive power and benefits, sustaining a resources base and demand for human resources appropriation. While agriculture and food systems attain SDGs 1 and 2, this should, however, not undermine achieving SDG 13, especially in developing countries [1]. In achieving SDG 13, various actions are required within the agricultural sector to reduce emissions, especially given that agriculture and food account for 24% of global CO_2 emissions and 10–25% of annual greenhouse gas (GHG) emissions, with livestock being the primary perpetrator. Conversely, while being a major GHG contributor, agriculture is also affected by climate extremes (the direct consequence of GHG emissions), which may exceed critical thresholds for crop and livestock production. For instance, it has been forecasted that by 2030, crop yield will decrease by 10–50% due to climate change [1]. This has a negative effect on the attainment of other SDGs such as SDG 1 and 2, which refer to poverty reduction and food security, respectively, in developing countries. One of the methods for reducing the impacts of climate change is to reduce the emissions from agricultural production.

The contribution of agriculture to South Africa's GDP has been decreasing since 1960, from over 10% to just above 2% in 2018. This can be explained by the economic transformation of the country from reliance on primary industries such as agriculture and mining, to manufacturing and services [8]. This has also been reflected in the overall electricity consumption of agriculture relative to other industries at 3%, with the sector contributing 7% to total GHG emissions in the country [9,10]. The agricultural sector is significant to South Africa's GDP and employment, as well as its GHG emissions and reductions [11]. The sector employs 661,000 people, representing 5% of all employment. One-tenth of these employees are labourers, whilst the rest are skilled workers [10].

Climate-change-related initiatives in South Africa's agricultural sector have embarked on an integration of climate smart agriculture into climate-resilient rural development [11]. However, the policy response to the nexus between energy use and productivity within the agricultural sector has been lacklustre. Some of the drivers and challenges of agricultural energy use in South Africa will include population increase, growing energy demand, intensification of energy use and economic growth (Table 1).

Table 1. Drivers, trends and challenges to energy use in South Africa.

Drivers	Key Trends	Future Challenges
Population increase and urbanization	Increase in the amount of energy use and energy in food production [12]	-Maintaining energy use whilst increasing food production
Growing energy demand	Increased energy use in agriculture, manufacturing, households, etc. [13,14]	-Providing adequate energy to agriculture without increasing pollution -Competing interest in terms of energy use between agriculture and other sectors of the economy
Increase in the amount of energy use in food production	Increased energy use in agricultural and manufacturing sectors [15,16]	-Ensuring sufficient, reliable and efficient energy for agriculture
Economic growth, industrialization and urbanization	Increasing non-renewable energy importation [12,17–20]	-Ensuring stable and quality energy supply for food production -Promoting private sector involvement in renewable energy utilisation for food production

With competing demands for resources and increasing environmental pressure, the challenge facing South Africa is how to minimize and manage conflicts between renewable and non-renewable energy uses for agricultural production given the commitment to reducing emissions and achieving the SDGs. The other problem is a lack of policy synergies between energy, agriculture and climate change in South Africa. Cross-sectoral efforts have remained linear, either taking into account the emissions from energy, or energy for agriculture. However, these relationships are dynamic, as shown in Figure 1.

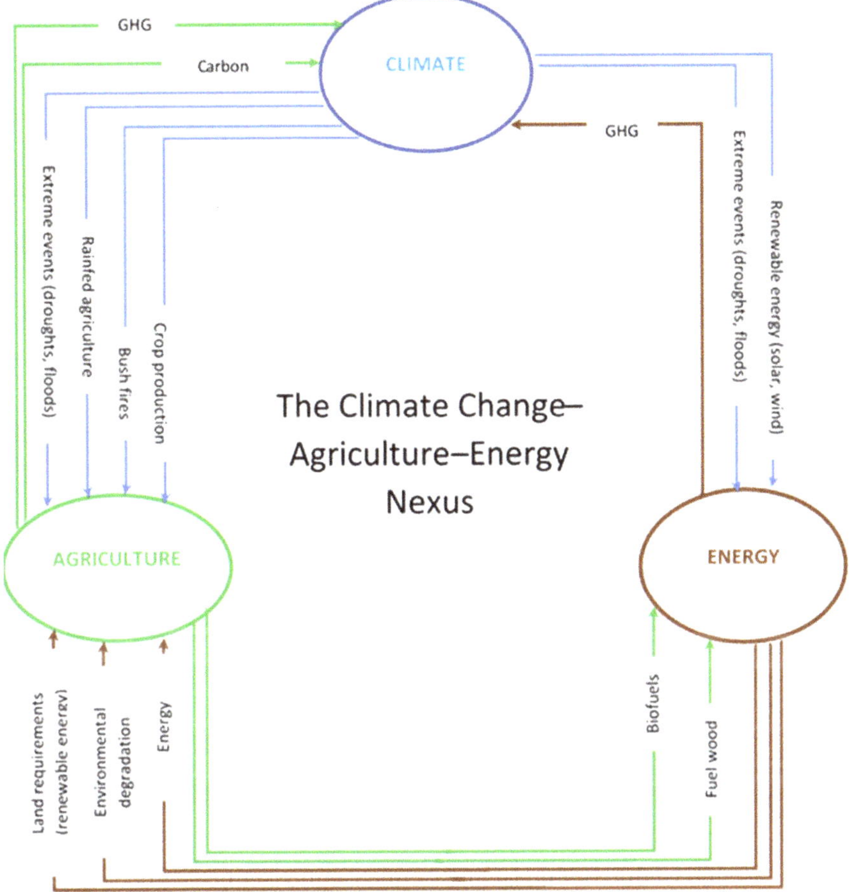

Figure 1. The climate change-agriculture-energy nexus. Adapted from Sridharan et al. [3].

1.1. Energy, Emissions and the Agricultural Sector

Globally, energy in the agricultural sector is mainly from fossil fuels [3,21]. Primary agricultural production tends to use between 17% and 20% of total energy globally. Between one-fifth and one-quarter of this is for primary farm production, whilst the rest is for post-harvest food operations. Total global agricultural energy consumption is around 6 EJ/year (expected to increase to 9 EJ/year by 2035), with only 1 EJ/year coming from renewable energy sources [3]. This energy is mainly used for irrigation, harvesting, livestock housing, heating and cultivation. Tractors, harvesters and machinery have utilised fossil fuels at a rate of between 11.1 GJ/ha and 20.4 GJ/ha [3]. Around 0.225 EJ/year is required globally to cover 324 million hectares of irrigated land, whilst 0.05 EJ/year is required for manufacturing and delivering irrigation equipment.

Despite the declining agricultural contribution to GDP, energy use in the South African agricultural sector has been consistent, as shown in Figure 2. Electricity and coal energy use in the sector peaked at 35,052, 8 TJ and 34,020 TJ in 1998 and 2012, respectively, with lows of 15,910, 2 TJ and 16,014, 4 TJ in 1993 and 2000, respectively [22]. According to Lin and Wesseh [13], energy directly affects spending decisions of firms and households, as well as economic performance. South Africa is an energy-intensive economy, with coal alone accounting for 72% of total primary energy consumption. This, however, is not reflected in the agriculture sector, with most coal consumption being exhibited in the energy and manufacturing sectors. The country accounts for 42% of the continent's CO_2 emissions and is the world's most carbon-intensive non-oil-producing developing country [13].

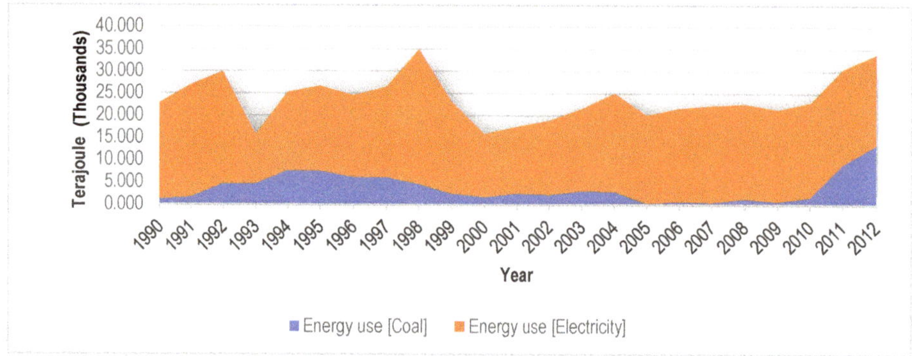

Figure 2. Energy use in South Africa's agriculture sector (the energy use does not take into account energy from methane and fuel oils, which contribute more than 50% of energy use in South Africa's agricultural sector [17]). Source: FAOSTAT [22].

South Africa is the world's 14th largest emitter of GHGs, mainly from coal [23]. At the 2015 Paris Agreement, nations undertook to cap emissions by 2025, at which level they will remain for a decade, and limit the increase in global temperatures to 1.5 °C by 2025. South Africa's nationally determined contributions (NDCs), each country's efforts at reducing national emissions and climate change adaptation, was determined at 2 °C. However, the country's commitments are highly insufficient. In the NDCs, South Africa pledged to shift away from coal and to end the expansion of nuclear power, whilst increasing renewable energy and gas [24,25]. South Africa's NDC is consistent with the Copenhagen Accord's proposed reduction of emissions by 34% in 2020 and 42% in 2025 [26]. A pertinent question that arises is what effect this emission (fossil fuels) reduction will have on agricultural production? What relationship exists between CO_2 emissions and agricultural production in South Africa? Not much has been documented concerning the relationship between the pledged emissions reductions and agricultural production in the country. Given that there are trade-offs between achieving the two objectives of emissions reduction and agricultural production (i.e., SDG 13 vs. SDG 1 and 2) [1], the objective of the study is to model how achieving SDGs 1 and 2 may affect the achievement of SDG 13 in South Africa. This is through predicting how poverty reduction and food security (i.e., an increase in agricultural income) might be achieved through cognizance of climate change actions.

Figure 3 shows the CO_2 emissions from agriculture in South Africa from 1990 to 2012. It is shown that most of the CO_2 emissions within the sector are from manure left on pastures, followed by electricity emissions and synthetic fertilisers [27]. Livestock and livestock products contribute 46–51% of agricultural income in South Africa, and have a large carbon footprint, mainly from land use and its changes (deforestation, feed production), as well as methane production [10,28,29]. The country's main agricultural activity (livestock production) induces the most environmental degradation from the sector. Contemplating the reduction of CO_2 emissions in the agricultural sector in South Africa requires climate change policy action targeting livestock production. Given that 40% of livestock farmers are smallholder farmers [30] and thus reliant on livestock production for food security and poverty reduction, the subsector exhibits trade-offs in achieving SDGs 1, 2 and 13.

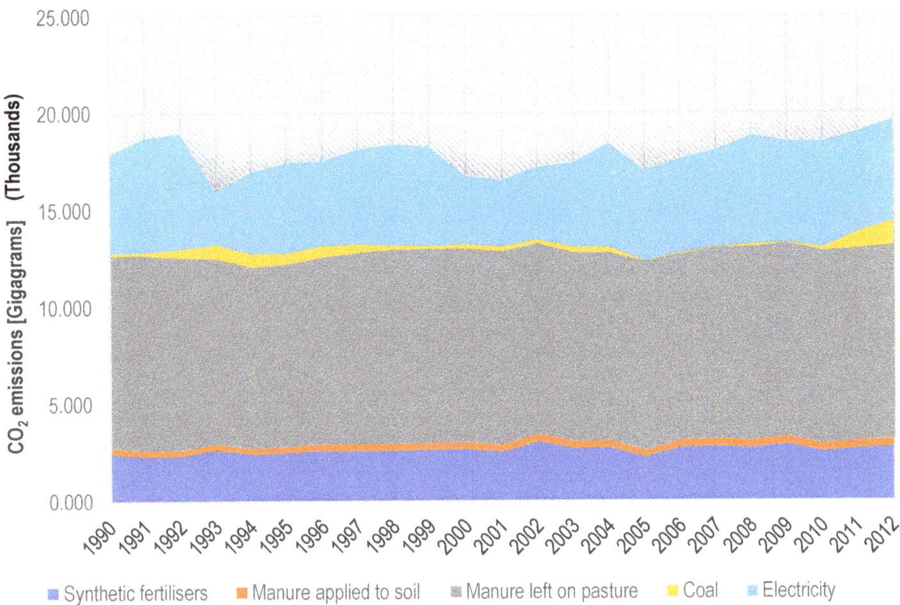

Figure 3. Carbon dioxide emissions from the South African agricultural sector. Source: FAOSTAT [27].

1.2. Conceptual Framework: Environmental Kuznets Curves

According to Bo [31], environmental quality indicators improve with an increase in income. However, there is a need for deeper investigation of the relationship between income and the environment. There is also a need to use data involving similar environmental involvement paths. It has also been shown that, sometimes, it is not always the case that environmental indicators improve with an increase in income, this being dependent on the indicators chosen. The objective of the study is to show the nexus between CO_2 emissions and agricultural production in South Africa through environmental Kuznets curves (EKC). Environmental Kuznets curves, in the context of this study, represent a temporal relationship between agricultural GDP and climate change (emissions). This offers the advantage of tracking the likelihood of achieving SDGs 1, 2 and 13 separately, and observing how they relate to each other. This is because the EKCs depict trade-offs between environmental degradation and economic growth. Economic growth can be exhibited through food security and poverty reduction. There has been little research done in South Africa utilising EKCs, with most concentrating on developed countries. The nature of an EKC makes it a challenge to apply the test in a low- or middle-income country. Furthermore, most studies utilising the EKC focus on relationship between GDP as an economic indicator and CO_2 as an environmental degradation indicator at the macrolevel [13,14,32]. Shahbaz, Kumar Tiwari, and Nasir [33], as well as Inglesi-Lotz and Bohlmann [34], utilised EKC in South Africa, concentrating on the overall economic development and growth in light of CO_2 emissions, with the studies not being sectoral-based. There is a lack of studies that are sectoral-focussed, for example in agriculture as an economic indicator [35–38]. Furthermore, Kijima et al. [32] highlighted that the inverted U-shape Kuznets curve hypothesis does not always hold, depending on the country in question, the variables used and the time period considered. The study provides a way to determine the relationship between the variables used in the EKC and the agricultural sectoral level in South Africa, as intended SDG indicators. The EKC should be a precursor for developing nations to pursue economic (sectoral) growth instead of implementing pro-environment policies [39]. Economic (sectoral) growth eventually leads to attaining both environmental and economic goals, whilst pro-environment policies slow down the economic (sectoral) growth.

The conceptual framework of the study is based on environmental Kuznets curves (EKC). EKCs increase awareness of environmental changes including global warming and climate change [39]. Environmental Kuznets curves (EKC) often reflect people paying more attention to environmental issues and resolving them with the help of increasing income. Thus, while the environmental quality initially gets worse, it then improves with economic development [32]. This is based on two perspectives: (i) more attention being given to quality of life as income increases, including better environmental protection and a healthy level of consumption. The government may eventually intervene for environmental protection, improving environmental quality; and (ii) the interaction of scale, structure and technology. On the one hand, an increased scale of production induces more energy consumption and increased environmental degradation. This is also exhibited by structural changes, but to a lesser and gradually decreasing extent. On the other hand, technology and R&D then tend to improve efficiency in energy use, thereby improving environmental quality [31,32]. Thus, it follows the sequence in Figure 4. A further school of thought argues that, instead of the typical inverted U-shaped EKC, it is actually an inverted N-shape, reflecting an initial decrease in environmental degradation, followed by an increase, with an eventual decrease when the economy is developing [40].

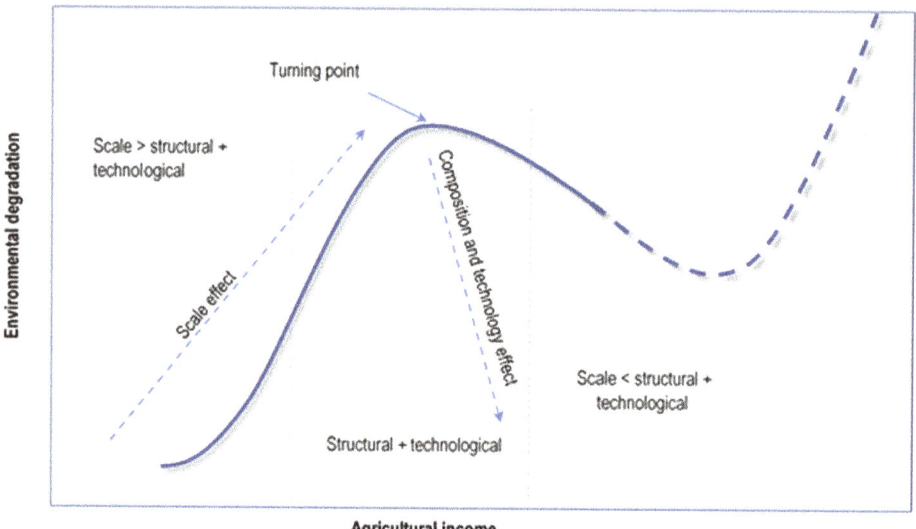

Figure 4. Kuznet curves model.

The study will take, as its focus, the interaction of scale, structure and technology. South Africa's agricultural sector is more developed than other SSA countries' in terms of agricultural production and poverty reduction. This is due to the relatively large-scale and technologically advanced agricultural sector. Even though authors such as Kaika and Zervas [41] highlight that Kuznets curves have been applied in economic transformation from primary production, to secondary and tertiary, with an associated initial increase, followed by stabilisation and then a decline in degradation of the environment, the current study argues that such an approach can be taken into the agricultural sector. The focus will be on the stage of development of the agricultural sector, which in itself depicts the level of environmental degradation. If the Kuznets theory proves otherwise, it reflects that improvements within the sector have not brought (or are yet to bring) about reductions in environmental degradation. It therefore reflects that, going forward, focussing on SDGs 1 and 2 would tend to reduce the impact of SDG 13. If the Kuznets theory is exhibited, then achieving SGDs 1, 2 and 13 is possible in the inverted U-shaped model, whilst at a later stage it will be reversed in the N-shaped model.

Empirical analysis of EKCs has centred upon gas indicators harmful to people's health like CO_2, CO, NO, SO_2, and the inverted U-shape was confirmed [31]. Other studies actually confirmed the inverted N-shape EKC. Some of the more recent studies on EKC are exhibited in Table 2. At the macro level, concentrating on overall GDP, EKCs have been exhibited in some studies, but not in others.

Table 2. Recent studies on environmental Kuznets curves.

Author	Period	Country/Region/Organization	Methodology	Variables Used in the Study	EKC Hypothesis
Balaguer and Cantavella [42]	1874–2011	Spain	Autoregressive distributed lag (ARDL) bounds test approach and error correction model (ECM)	Per capita CO_2, GDP, crude oil prices	Exhibited
Alam, Murad, Noman, and Ozturk [43]	1970–2012	Brazil, China, India and Indonesia	ARDL and ECM	Per capita CO_2, GDP, energy, Trade openness	Exhibited in India, but not in Brazil, China and Indonesia
Apergis [44]	1960–2013	15 OECD countries	Common correlated effects and panel quantile cointegration test	Emissions, per capita GDP	Mixed results
Al-Mulali and Ozturk [45]	1990–2012	27 Countries	Kao and Fisher cointegration and VECM	CO_2, GDP, renewable energy consumption, non-renewable energy consumption, trade, population, energy prices	Exhibited
Ahmad et al. [46]	1992–2011	Croatia	ARDL and VECM	CO_2, GDP	Exhibited
Özokcu and Özdemir [47]	1980–2010	26 OECD countries and 52 emerging countries	Polynomial (cubic) regression model	CO_2 per capita, GDP per capita, energy use per capita	Mixed results
Churchill, Inekwe, Ivanovski, and Smyth [48]	1870–2014	20 OECD countries	Panel cointegration, mean group estimator (MGE), common corelated mean group (CCEMG), augmented mean group (AMG) and pooled MG (PMG) estimator	CO_2, GDP, trade, population, financial development	Mixed results

2. Materials and Methods

The study assessed the impact of agricultural production on agricultural CO_2 emissions in South Africa. The study utilised time series data for the period 1990 to 2012; even though agricultural income data, as well as national CO_2 emissions data, were available for South Africa up to 2018, sectoral-based CO_2 emissions data were only available up to 2012, which placed a limitation on the time series dataset. The time series data were indexed to the 2004 constant figures. The data were collected from FAOSTAT and the World Bank [22,27,49]. The variables utilised included agricultural carbon dioxide emissions (CO_2) as the dependent variable, with agricultural GDP, agricultural coal energy consumption and agricultural electricity energy consumption as explanatory variables. The agricultural CO_2 utilised combined CO_2 from fertiliser, manure applied in soil, manure left on pastures, coal and electricity.

The study utilised the method as utilised by Baek [40] as well as He and Richard [50] in estimating three models: log linear, log quadratic and log cubic:

$$E = F(Y, Z) \tag{1}$$

$$E = F\left(Y, Y^2, Z\right) \tag{2}$$

$$E = F\left(Y, Y^2, Y^3, Z\right), \tag{3}$$

where E is the CO_2 emissions from agricultural activities, Y is the agricultural GDP and Z is another explanatory variable that influences environmental degradation in agricultural production. The main objective of the study was establishing a relationship between agricultural income and agricultural CO_2 emissions. The estimated models in logarithmic form are as follows:

$$\ln(E)_t = \beta_0 + \beta_1 \ln Y_t + \beta_4 \ln ACC_t + \beta_5 \ln AEC_t + \varepsilon_t \tag{4}$$

$$\ln(E)_t = \beta_0 + \beta_1 \ln Y_t + \beta_2 (\ln Y_t)^2 + \beta_4 \ln ACC_t + \beta_5 \ln AEC_t + \varepsilon_t \tag{5}$$

$$\ln(E)_t = \beta_0 + \beta_1 \ln Y_t + \beta_2 (\ln Y_t)^2 + \beta_3 (\ln Y_t)^3 + \beta_4 \ln ACC_t + \beta_5 \ln AEC_t + \varepsilon_t \tag{6}$$

where t is the time period, E is the CO_2 emissions from agricultural activities, Y is the real agricultural income, ACC is the agricultural coal energy consumption, AEC is the agricultural electricity consumption and ε_t is the standard error term. The elasticity of CO_2 with respect to agricultural income in the typical inverse U-shaped EKC form should be positive ($\beta_1 > 0$), whilst the agricultural income elasticity of its square would be negative ($\beta_2 < 0$). The agricultural income elasticity of its cubic would be positive ($\beta_3 > 0$) for the N-shape EKC hypothesis to be true. Agricultural coal and electricity energy consumption elasticities would be expected to be positive ($\beta_4\beta_5$), meaning that higher coal and electricity consumption will result in higher CO_2 emissions in the agricultural sector.

The study utilised the autoregressive distributive lag (ARDL) bounds test as well as the error correction model (ECM) for estimating the long-run adjustment process toward equilibrium [51,52]. This method is advantageous in that regressions can be carried out regardless of integration of I (1) or I (0), with most macroeconomic variables being either of these two orders. Another advantage is that serial correlation and endogeneity problems are removed when simultaneously taking appropriate long-run and short-run lags.

The relationships among agricultural CO_2, agricultural income, agricultural coal consumption and agricultural electricity consumption in Equations (4)–(6) follow a time path before long-run nexus is achieved. Thus, Equations (4)–(6) would be written as an unrestricted error correction specification:

$$\Delta \ln CO_{2t} = \alpha_0 + \sum_{i=1}^{p} \alpha_i \Delta \ln CO_{2t-i} + \sum_{i=1}^{p} \varphi_i \Delta \ln y_{t-i} + \sum_{i=1}^{p} \delta_i \Delta \ln ACC_{t-i} +$$
$$\sum_{i=1}^{p} \omega_i \Delta \ln AEC_{t-i} + \lambda_1 \ln CO_{2t-1} + \lambda_2 \ln y_{t-1} + \lambda_4 \ln ACC_{t-1} + \lambda_5 \ln AEC_{t-1} + \varepsilon_t \tag{7}$$

$$\Delta \ln CO_{2t} = \alpha_0 + \sum_{i=1}^{p} \alpha_i \Delta \ln CO_{2t-i} + \sum_{i=1}^{p} \varphi_i \Delta \ln y_{t-i} + \sum_{i=1}^{p} \gamma_i \Delta (\ln y_{t-i})^2 +$$
$$\sum_{i=1}^{p} \delta_i \Delta \ln ACC_{t-i} + \sum_{i=1}^{p} \omega_i \Delta \ln AEC_{t-i} + \lambda_1 \ln CO_{2t-1} + \lambda_2 \ln y_{t-1} + \lambda_3 (\ln y_{t-1})^2 +$$
$$\lambda_4 \ln ACC_{t-1} + \lambda_5 \ln AEC_{t-1} + \varepsilon_t \tag{8}$$

$$\Delta \ln CO_{2t} = \alpha_0 + \sum_{i=1}^{p} \alpha_i \Delta \ln CO_{2t-i} + \sum_{i=1}^{p} \varphi_i \Delta \ln y_{t-i} + \sum_{i=1}^{p} \gamma_i \Delta (\ln y_{t-i})^2 +$$
$$\sum_{i=1}^{p} \delta_i \Delta \ln ACC_{t-i} + \sum_{i=1}^{p} \omega_i \Delta \ln AEC_{t-i} + \sum_{i=1}^{p} \varnothing_i \Delta (\ln y_{t-i})^3 + \lambda_1 \ln CO_{2t-1} +$$
$$\lambda_2 \ln y_{t-1} + \lambda_3 (\ln y_{t-1})^2 + \lambda_4 \ln ACC_{t-1} + \lambda_5 \ln AEC_{t-1} + \lambda_6 (\ln y_{t-1})^3 + \varepsilon_t \tag{9}$$

where ε_t are the new serially independent errors. The estimation procedure initially tests whether there is evidence of a cointegration relationship through the ARDL bounds test. The null hypothesis of no cointegration ($H_0 : \lambda_1 = \lambda_2 = \lambda_3 = \lambda_4 = \lambda_5 = \lambda_6 = 0$) is tested against the alternative hypothesis ($H_1 : \lambda_1 \neq \lambda_2 \neq \lambda_3 \neq \lambda_4 \neq \lambda_5 \neq \lambda_6 \neq 0$). If the F-statistics from the ordinary least squares go beyond the upper bound of the critical values provided in Pesaran et al. [51] and Pesaran and Shin [53], then the null hypothesis is rejected, exhibiting a cointegrating relationship among the variables. If the F-statistics are below the lower bound, we fail to reject the null hypothesis. When the F-statistics lie between the upper and lower critical values, the test results will be inconclusive. The next stage will be to estimate long-run coefficients of the cointegrating relation and make inferences.

The final stage will involve estimating an error correction model, taking the form of Equations (10)–(12) but including the long-run terms in the error correction variable lagged one period:

$$\Delta \ln CO_{2t} = \alpha_0 + \sum_{i=1}^{p} \alpha_i \Delta \ln CO_{2t-i} + \sum_{i=1}^{p} \varphi_i \Delta \ln y_{t-i} + \sum_{i=1}^{p} \delta_i \Delta \ln ACC_{t-i} + \\ \sum_{i=1}^{p} \omega_i \Delta \ln AEC_{t-i} + \lambda ect_{t-1} + \varepsilon_t \tag{10}$$

$$\Delta \ln CO_{2t} = \alpha_0 + \sum_{i=1}^{p} \alpha_i \Delta \ln CO_{2t-i} + \sum_{i=1}^{p} \varphi_i \Delta \ln y_{t-i} + \sum_{i=1}^{p} \gamma_i \Delta (\ln y_{t-i})^2 + \\ \sum_{i=1}^{p} \delta_i \Delta \ln ACC_{t-i} + \sum_{i=1}^{p} \omega_i \Delta \ln AEC_{t-i} + \lambda ect_{t-1} + \varepsilon_t \tag{11}$$

$$\Delta \ln CO_{2t} = \alpha_0 + \sum_{i=1}^{p} \alpha_i \Delta \ln CO_{2t-i} + \sum_{i=1}^{p} \varphi_i \Delta \ln y_{t-i} + \sum_{i=1}^{p} \gamma_i \Delta (\ln y_{t-i})^2 + \\ \sum_{i=1}^{p} \delta_i \Delta \ln ACC_{t-i} + \sum_{i=1}^{p} \omega_i \Delta \ln AEC_{t-i} + \sum_{i=1}^{p} \varnothing_i \Delta (\ln y_{t-i})^3 + \lambda ect_{t-1} + \varepsilon_t \tag{12}$$

where ect_{t-1} is the error correction term represented by the OLS residual series from the long-run cointegration relationship, and the λ coefficient indicates the speed of adjustments towards this long-run equilibrium. Diagnostic tests such as the Breusch-Godfrey serial correlation LM test, the Jarque-Bera Test and the Breusch-Pagan-Godfrey Test were used to test for collinearity, normality and heteroscedasticity, respectively. Cumulative sum (CUSUM) and cumulative sum of squares (CUSUMSQ) were used to test the long- and short-run stability of the model.

3. Results and Discussion

3.1. Descriptive Results

Table 3 shows that the average agricultural CO_2 emissions for the 22-year period was 17,935.49 gigagrams, with a maximum of 19,093.30 gigagrams and a minimum of 15,993.25 gigagrams. The value of agricultural production has averaged R 69.17 billion, whilst coal and electricity consumption within the sector had means of 3480.81 kilojoules and 20,190.75 kilojoules, respectively. The results indicate a strong positive correlation between agricultural CO_2 emissions and the value of agriculture, as well as electric energy use.

Table 3. Descriptive statistics.

	CO$_2$ Emissions (Gigagrams)	Gross Value of Agriculture (R Million)	Coal Energy (Kilojoules)	Electricity Energy (Kilojoules)
Mean	17,935.49	69,168.65	3840.809	20,190.75
Median	18,093.30	52,185.60	2605.800	20,718.00
Maximum	19,646.07	16,8591.1	13,467.60	30,357.20
Minimum	15,993.25	20,198.00	361.0000	11,188.80
Std Dev.	899.9911	44,950.37	3294.036	3915.104
Skewness	−0.246352	0.790601	1.260109	0.133680
Kurtosis	2.446791	2.393959	4.221092	4.191326
Correlation				
CO$_2$ emissions	1.000			
Gross value of agriculture	0.525	1.000		
Coal energy	0.264	0.142	1.000	
Electricity energy	0.747	0.095	−0.030	1.000

3.2. Empirical Results

Testing unit roots' properties is necessary when applying any standard cointegration in examining the long-run relationship between variables. The augmented Dickey-Fuller (ADF) test created by Said and Dickey [54] was used to test for unit roots. The series used in the analysis had a unit root problem. Table 4 shows that the series are integrated at different orders because AEC is of I (0), whilst the rest are I (1). The ARDL bounds test was therefore necessary for establishing the long-run relationship.

Table 4. Augmented Dickey-Fuller unit root test.

	ADF Statistics I (0)	ADF Statistics I (1)
$\ln CO_{2t}$	−2.48	−5.44 ***
$\ln Y_t$	−2.33	−5.27 ***
$(\ln Y_t)^2$	0.81	−5.24 ***
$(\ln Y_t)^3$	1.68	−4.87 ***
$\ln ACC_t$	−1.61	−4.66 ***
$\ln AEC_t$	−3.78 **	
Critical values	1%	−3.809
	5%	−3.021
	10%	−2.650

Max lag = 2; Schwarz info. criterion; Sig at ** 5%, *** 1%.

Selecting an appropriate lag length is necessary for applying the ARDL bounds testing approach to cointegration. Table 5 shows the lag selection criteria used in the study. The selection criteria was based on the Akaike information criterion (AIC) and Schwarz information criterion (SIC) statistics. The AIC is superior for small sample datasets [33]. An appropriate lag length can be used to capture dynamic linkages between series [55]. The maximum lag selections for the dependent and regressor values was 1.

Table 5. Lag selection criterion.

	AIC			SC		
Lag	0	1	2	0	1	2
$\ln CO_{2t}$	−3.03	−3.12 *	−3.04	−2.98	−3.02 *	−2.89
$\ln Y_t$	1.92	−1.98 *	−1.9	1.97	−1.89 *	−1.75
$(\ln Y_t)^2$	6.29	2.48 *	2.56	6.34	2.58 *	2.70
$(\ln Y_t)^3$	10.12	6.41 *	6.48	10.17	6.51 *	6.63
$\ln ACC_t$	2.89	2.38 *	2.46	2.94	2.48 *	2.61
$\ln AEC_t$	−0.22 *	−0.22	−0.22	−0.22 *	−0.12	−0.07

Sig at * 5%.

Table 6 shows the results of the ARDL bounds test approach and short-run results. The results show that, for the linear, squared and cubic models, the F-statistic was greater than the critical value of the upper bound I (1), and thus there is cointegration and a long-run relationship in each of the models. In the linear model, the previous period's CO_2 emissions levels, agricultural income, agricultural coal consumption and agricultural electricity consumption contribute to the CO_2 emissions levels in the next period. The table shows that a 1% increase in the previous period's CO_2 emissions causes a 0.28% increase in the next period's CO_2 emissions. A 1% increase in the agricultural income, coal consumption and electricity consumption within the sector induces a 0.023%, 0.022% and 0.18% increase in CO_2 emissions, respectively.

For the squared agricultural income model, however, the linear agricultural income had a negative coefficient, whilst the squared agricultural income had a positive coefficient. In the squared agricultural income model, a 1% increase in agricultural income and the square of agricultural income induced a 0.35% reduction and 0.044% increase in CO_2, respectively. This is mainly based on the characteristics within the sector, which is livestock-based, accounting for 46–51% of agricultural GDP [28]. Livestock has a large carbon footprint. Increasing income within the agricultural sector means increasing livestock production and productivity. This therefore results in increases of CO_2 from manure on pastures, for instance, which is the largest CO_2 emitter within the sector.

Table 6. Bounds test and short-run relationship.

Variable	/		∧		∿	
$\ln CO_{2t}\,(-1)$	0.28 (2.91) **		−0.25 (−1.31)		−0.26 (−1.24)	
$\ln Y_t$	0.023 (2.97) ***		−0.35 (−3.18) ***		−0.20 (−0.12)	
$(\ln Y_t)^2$			0.044 (3.46) ***		0.0079 (0.019)	
$(\ln Y_t)^3$					0.0028 (0.090)	
$\ln ACC_t$	0.022 (3.72) ***		0.014 (3.26) ***		0.013 (2.06) *	
$\ln ACC_t\,(-1)$	−0.011 (−1.67)					
$\ln AEC_t$	0.18 (8.28) ***		0.16 (9.03) ***		0.16 (8.61) ***	
$\ln AEC_t\,(-1)$			0.087 (2.22) **		0.088 (2.14) *	
C			5.20 (5.95) ***		5.01 (2.13) *1	
R-squared	0.887683		0.930861		0.930900	
Adjusted R-squared	0.852584		0.903205		0.896351	
F-statistic	25.29077 ***		33.65891 ***		26.94370 ***	
Durbin-Watson statistic	2.434274		2.214419		2.210267	
F-bounds test						
F-statistic	28.11		11.69		9.09	
	I (0)	I (1)	I (0)	I (1)	I (0)	I (1)
10%	2.72	3.77	2.45	3.52	2.26	3.35
5%	3.23	4.35	2.86	4.01	2.62	3.79
2.5%	3.69	4.89	3.25	4.49	2.96	4.18
1%	4.29	5.61	3.74	5.06	3.41	4.68

Sig. at * 10%, ** 5%, *** 1%; *t*-values in parentheses.

After examining the cointegration of the variables, the next stage was disclosing the impact of agricultural income, agricultural coal energy consumption and agricultural electricity energy consumption on agricultural CO_2 emissions in the long run. The EKC hypothesis was not validated in the long run. The squared model showed that the linear agricultural income was negative, whilst the quadratic coefficients was positive, which is contrary to the inverted U-shaped EKC hypothesis. The cubic model also shows the linear agricultural income coefficient being negative, whilst both the quadratic and cubic coefficients were positive, which is also contrary to the inverted N-shaped EKC postulations. It is therefore deduced that the EKC hypothesis is not exhibited within the South African agricultural sector based on agricultural CO_2 emissions. The results fall short of Shahbaz, Aviral, and Nasir [56] and Shahbaz et al. [33], who reported a validation of the EKC in South Africa. The differences can be explained by the scale of their studies, which were focussing on the overall economy and not sector-based. Furthermore, their studies utilised a longer time series (1965–2008), which was able to capture economic transformations over a period of time. However, Inglesi-Lotz and Bohlmann [34] also utilised a long time series (1960–2010), the same as Nasr, Gupta, and Sato [57] (1911–2010), and could not validate the EKC in South Africa. It was highlighted that the economy was still transforming through the early stages of the inverted U-shape EKC. Table 7 shows that in the linear model, agricultural income, coal energy consumption and electricity energy consumption had a significant, positive impact on agricultural CO_2 emissions. In the squared model, agricultural income had a significant, negative impact on agricultural CO_2 emissions, whilst the squared agricultural income, agricultural coal energy consumption and agricultural electricity energy consumption had a significant, positive influence on the agricultural CO_2 emissions. For the cubic model, agricultural income had no influence on the agricultural CO_2 emissions (whether linear, squared or cubic), whilst agricultural coal energy consumption and agricultural electricity energy consumption had a significant, positive influence on the agricultural CO_2 emissions.

Table 7. Error correction model and long-run relationship.

Variable	╱	⌃	∿
$\ln Y_t$	0.032 (3.06) ***	−0.28 (−3.58) ***	−0.12 (−0.11)
$(\ln Y_t)^2$		0.035 (3.95) ***	0.0063 (0.019)
$(\ln Y_t)^3$			0.0022 (0.090)
$\ln ACC_t$	0.015 (2.00) *	0.011 (3.12) ***	0.011 (1.9) *
$\ln AEC_t$	0.25 (5.97) ***	0.20 (10.06) ***	0.20 (9.70) ***
CointEq (-1)	−0.719 ***	−1.247 ***	−1.253 ***

Sig. at * 10%, ** 5%, *** 1%; *t*-values in parentheses.

Table 7 shows that a 1% increase in the agricultural income induces a 0.03% increase and 0.28% decrease in the agricultural CO_2 emissions in the linear and quadratic models, respectively. Increasing the squared agricultural income will induce a 0.035% increase in the CO_2 emissions. This shows that expanding the sector is not having a significant effect in terms of CO_2 emissions reduction. Thus, food security and poverty reduction (by way of agricultural income growth) are increasing CO_2 emissions. There is thus a trade-off between achieving agricultural income growth (SDG 1 and 2) and reducing CO_2 emissions (SDG 13). In all three models it was shown that a 1% increase in the coal energy consumption in agriculture will induce a 0.011-0.015% increase in agricultural CO_2 emissions, whilst a 1% increase in agricultural electricity consumption will result in a 0.20-0.25% increase in agricultural CO_2 emissions. This shows that, going into the future, electricity consumption will account for a large portion of the emissions from the agricultural sector. The findings were, however, different in Shahbaz et al. [33], who found that coal consumption was the major source of CO_2 emissions within the South African economy. This was mainly based on the scale of the study: they concentrated on the whole economy and not a specific sector. Even though South Africa is the sixth-largest consumer of coal, its use appears to be primarily in the energy and manufacturing sectors, and not necessarily in the agricultural sector [9]. The Error Correction Term (ECT) has a negative and significant sign at the 1% level, indicating a long-run association between the variables. The ECT from the linear model shows that the change in CO_2 emissions from the short run towards the long run is estimated at 71.9% every year in the linear model, whilst it is 124.7% and 125.3% for the squared and cubic models.

A pairwise Granger causality test was also performed to analyse the direction of causal relationship with regards to CO_2 emissions. The causality test was necessary for establishing whether agricultural growth causes environmental degradation, or whether the emissions from agricultural activities were responsible for the sector's growth. The pairwise Granger causality test in Table 8 shows that agricultural income, as well as its squared and cubic terms, tend to Granger cause CO_2 emissions within the sector in the long run at the 1% level. Even though this was at the overall GDP level, this finding is consistent with Shahbaz et al. [33] in South Africa. However, coal and electric energy do not Granger cause CO_2 emissions.

Table 8. Pairwise Granger causality test.

	F-Statistic	Prob.
$\ln Y_t$ does not Granger cause $\ln CO_{2t}$	3.15122	0.0702
$(\ln Y_t)^2$ does not Granger cause $\ln CO_{2t}$	3.07631	0.0741
$(\ln Y_t)^3$ does not Granger cause $\ln CO_{2t}$	3.11833	0.0718
$\ln ACC_t$ does not Granger cause $\ln CO_{2t}$	0.57636	0.5732
$\ln AEC_t$ does not Granger cause $\ln CO_{2t}$	0.86314	0.4406

3.3. Diagnostic Tests

A sensitivity analysis indicates that the models pass all diagnostic tests. Table 9 shows the collinearity, heteroskedasticity and normality diagnostic tests. The collinearity test in all three instances shows an insignificant value, indicating no collinearity. There was also no heteroskedasticity, and normal distribution for the residual within the models. The models were thus robust.

Table 9. Residual diagnostic tests.

	F-Stat	Prob.	F-Stat	Prob.	F-Stat	Prob.
Breusch-Godfrey serial correlation LM test	2.208748	0.1467	0.711048	0.5093	0.948005	0.4147
Breusch-Pagan-Godfrey heteroskedasticity test	2.058414	0.1245	2.125251	0.1107	1.733281	0.1804
Jarque-Bera normality test	2.962189	0.227389	0.9898624	0.609685	0.987404	0.610363

The long-run parameter stability was tested by applying the CUSUM and CUSUM of square test. Figures 5–7 show the CUSUM and CUSUM of square stability tests for the models. All the models lie within the critical bounds, and therefore confirm the stability of long-run estimates.

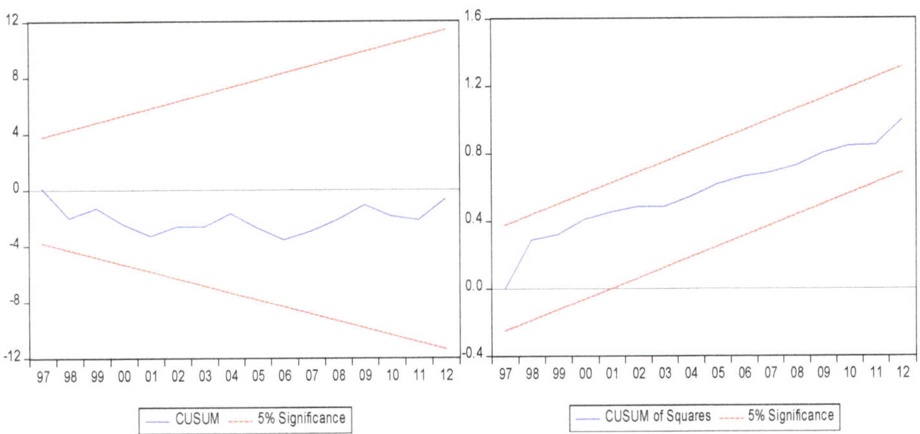

Figure 5. CUSUM and CUSUM of squares test for the linear model.

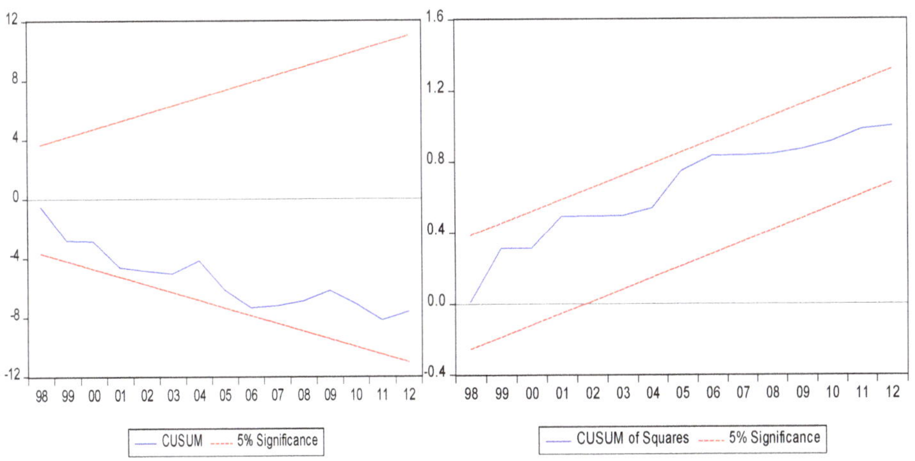

Figure 6. CUSUM and CUSUM of squares test for the squared model.

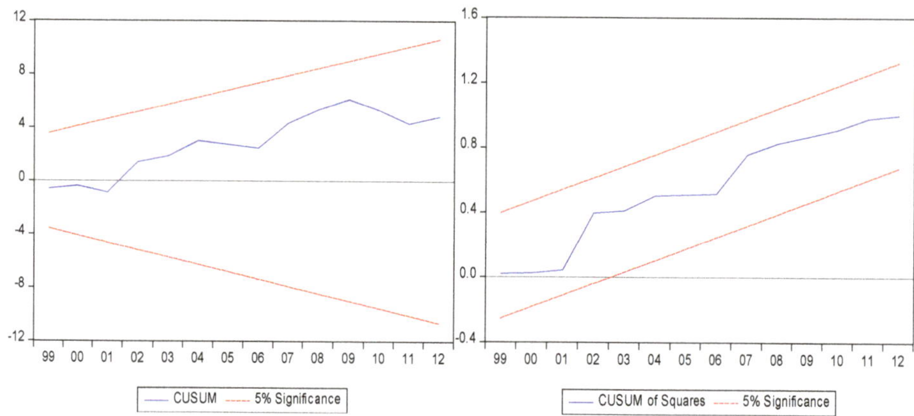

Figure 7. CUSUM and CUSUM of squares test for the cubic model.

4. Conclusions and Recommendations

The study sought to investigate the trade-offs between SDGs 1 (poverty reduction), 2 (food security) and 13 (climate change action) through analysing the validation of environmental Kuznets curves (EKC), which examine the nexus between income and environmental degradation. The study focused on the agricultural sector in South Africa for the period 1990 to 2012, and the income that was utilised in the study was real agricultural income (2004 constant levels), whilst the indicator used for environmental degradation was agricultural CO_2 emissions. The agricultural CO_2 variables used in the study include agricultural income, agricultural income squared, agricultural income cubic, agricultural coal consumption and agricultural electricity consumption. The bounds testing approach was used for examining the long-run relationship between the variables. The study finds cointegration between the series in the three models (linear, squared and cubic). Furthermore, agricultural coal energy and electricity energy consumption tend to increase agricultural CO_2 emissions in South Africa. The EKC hypothesis is not validated in the South African agricultural sector as there is a U-shaped EKC in the squared model, whilst no relationship was exhibited in the cubic model. Thus, as agricultural income increases, so do CO_2 emissions in the South African agricultural sector. This implies that for South Africa to reduce its CO_2 emissions and achieve SDG 13, the country has to sacrifice agricultural growth. This is not feasible as agriculture is required for food security and hunger reduction (SDG 2), as well as poverty reduction (SDG1). This further shows that the sector is less likely to be used to overcome environmental degradation, especially given the country's 2 °C nationally determined contributions (NDCs). This is also exacerbated by the focus of EKC, which disregards consumption evolution and focusses on production activity [35]. Thus, promoting sectoral growth, as implied by the EKC hypothesis, will not result in simultaneous attainment of both sectoral and environmental goals. In this instance, a pro-environmental policy would be ideal for reducing the emissions from the country's agricultural sector, as opposed to a pro-sector-growth one.

The study recommends that policies aimed at improving energy efficiency be promoted in order to decrease agricultural CO_2 emissions without adversely affecting the agricultural sector's productive capacity. To ensure the positive impact of agricultural coal energy and electricity energy on agricultural CO_2 emissions, there needs to be a policy of searching for and using renewable energy (wind, solar, biodiesel fuel) within the South African agricultural sector. There is large potential for bioenergy and hydroelectricity use in the South African agricultural sector, but it will require buy-in and investment [58]. Policy makers could also increase taxes on fossil fuel use within the agricultural sector, whilst subsidising renewable energies. Furthermore, instead of focusing entirely on policy intervention, the country could also invest more in R&D for more efficient technology to be used in the sector. This could aid in reducing CO_2 emissions. Ultimately, the study recommends that policy within the sector should not be linear, focussing on cause-effect, as suggested by the EKC hypothesis. There is a need for simultaneous growth and protection of the environment [39].

Author Contributions: Conceptualization, S.N.; methodology, J.A.; validation, J.A., L.Z. and S.T.; formal analysis, S.T.; investigation, S.N.; writing—original draft preparation, S.N.; writing—review and editing, J.A.; supervision, L.Z.; project administration, S.N.

Funding: This research received no external funding.

Acknowledgments: The authors acknowledge the valuable comments from the anonymous reviewers.

Conflicts of Interest: The authors declare no conflict of interest.

References

1. Campbell, B.M.; Hansen, J.; Rioux, J.; Stirling, C.M.; Twomlow, S.; Wollenberg, E.L. Urgent action to combat climate change and its impacts (SDG 13): Transforming agriculture and food systems. *Curr. Opin. Environ. Sustain.* **2018**, *34*, 13–20. [CrossRef]
2. United Nations. *Transforming Our World: The 2030 Agenda for Sustainable Development*; United Nations: Geneva, Switzerland, 2015; Volume 16301.
3. Sridharan, V.; Howells, M.; Ramos, E.P.; Fuso-Nerini, F.; Sundin, C.; Almulla, Y. The Climate-Land-Energy and Water Nexus: Implications for agricultural research. In Proceedings of the Science Forum 2018, Stellenbosch, South Africa, 10–12 October 2018; pp. 1–57.
4. UNFCCC. *United Nations Framework Convention on Climate Change (UNFCCC)*; Paris Agreement; UNFCCC: Paris, France, 2015; Report No.: Annex to decision 1/CP.21 document FCCC/CP/2015/10/Add.1; Available online: http://unfccc.int/resource/docs/2015/cop21/eng/10a01.pdf#page=24 (accessed on 23 June 2019).
5. IPCC. *Climate Change 2014: Synthesis Report: Fifth Assessment Report of the Intergovernmental Panel on Climate Change*; IPCC: Geneva, Switzerland, 2014; Available online: http://www.ipcc.ch/pdf/assessment-report/ar5/syr/SYR_AR5_SPM.pdf%0A6 (accessed on 19 June 2019).
6. Canavan, C.R.; Graybill, L.; Fawzi, W.; Kinabo, J. The SDGs will require integrated agriculture, nutrition, and health at the community level. *Food Nutr. Bull.* **2016**, *37*, 112–115. [CrossRef] [PubMed]
7. van Noordwijk, M.; Duguma, L.A.; Dewi, S.; Leimona, B.; Catacutan, D.C.; Lusiana, B.; Öborn, I.; Hairiah, K.; Minang, P.A. SDG synergy between agriculture and forestry in the food, energy, water and income nexus: Reinventing agroforestry? *Curr. Opin. Environ. Sustain.* **2018**, *34*, 33–42. [CrossRef]
8. Meyer, N.G.; Breitenbach, M.C.; Fenyes, T.I.; Jooste, A. The economic rationale for agricultural regeneration and rural infrastructure investment in South Africa. *J. Dev. Perspect.* **2007**, *3*, 73–83.
9. DEA. *South Africa's Greenhouse Gas Inventory Report 2000–2010*; DEA: Pretoria, South Africa, 2014. Available online: https://www.environment.gov.za/sites/default/files/docs/greenhousegas_invetorysouthafrica.pdf (accessed on 6 July 2019).
10. USAID. *Greenhouse Gas Emissions in South Africa*; USAID: Washington, DC, USA, 2014. Available online: https://www.environment.gov.za/sites/default/files/docs/greenhousegas_invetorysouthafrica.pdf%0A%0A (accessed on 12 December 2017).
11. CSIR. *Final Technical Report: Intended Nationally Determined Contributions*; CSIR: Pretoria, South Africa, 2015.
12. Pegels, A. Renewable energy in South Africa: Potentials, barriers and options for support. *Energy Policy* **2010**, *38*, 4945–4954. [CrossRef]
13. Lin, B.; Wesseh, P.K. Energy consumption and economic growth in South Africa reexamined: A nonparametric testing apporach. *Renew. Sustain. Energy. Rev.* **2014**, *40*, 840–850. [CrossRef]
14. Menyah, K.; Wolde-Rufael, Y. Energy consumption, pollutant emissions and economic growth in South Africa. *Energy Econ.* **2010**, *32*, 1374–1382. [CrossRef]
15. Koen, R.; Holloway, J.; Mokilane, P.; Makhanya, S.; Magadla, T. *Forecasts for Electricity Demand in South Africa (2017–2050) Using the CSIR Sectoral Regression Model for the Integrated Resource Plan of South Africa (as inputs into the Integrated Resource Plan)*; Department of Energy: Pretoria, South Africa, 2017. Available online: http://www.energy.gov.za/IRP/irp-update-draft-report2018/CSIR-annual-elec-demand-forecasts-IRP-2015.pdf (accessed on 30 July 2019).
16. Inglesi-Lotz, R.; Pouris, A. Energy efficiency in South Africa: A decomposition exercise. *Energy* **2012**, *42*, 113–120. [CrossRef]

17. Ratshomo, K.; Nembahe, R. *South African Energy Sector Report*; Department of Energy: Pretoria, South Africa, 2018; pp. 2–49. Available online: http://www.energy.gov.za (accessed on 5 July 2019).

18. Jain, S.; Jain, P.K. The rise of Renewable Energy implementation in South Africa. *Energy Procedia* **2017**, *143*, 721–726. [CrossRef]

19. Rafey, W.; Sovacool, B.K. Competing discourses of energy development: The implications of the Medupi coal-fired power plant in South Africa. *Glob. Environ. Chang.* **2011**, *21*, 1141–1151. [CrossRef]

20. Walwyn, D.R.; Brent, A.C. Renewable energy gathers steam in South Africa. *Renew. Sustain. Energy Rev.* **2015**, *41*, 390–401. [CrossRef]

21. Bundschuh, J.; Chen, G. *Sustainable Energy Solutions in Agriculture*; CRC Press: Boca Raton, FL, USA, 2014.

22. FAOSTAT. Agriculture Energy Consumption—South Africa. 2019. Available online: http://www.fao.org/faostat/en/#data/GT (accessed on 1 July 2019).

23. McSweeney, R.; Timperely, J. The Carbon Brief Profile—South Africa. 2019. Available online: https://www.carbonbrief.org/the-carbon-brief-profile-south-africa (accessed on 28 June 2019).

24. DEA. *South Africa's Intended Nationally Developed Contribution*; DEA: Pretoria, South Africa, 2015. Available online: https://www.environment.gov.za/sites/default/files/docs/sanational_determinedcontribution.pdf (accessed on 16 June 2019).

25. Climate Smart Tracker. South Africa—Country Summary. 2019. Available online: https://climateactiontracker.org/countries/south-africa/ (accessed on 27 June 2019).

26. StatsSA. *Sustainable Development Goals: Indicator Baseline Report 2017—South Africa*; Statistics South Africa: Pretoria, South Africa, 2017.

27. FAOSTAT. Agricultural Total Emission—South Africa 2019. Available online: http://www.fao.org/faostat/en/#data/GT (accessed on 1 July 2019).

28. DAFF. *Trends in the Agricultural Sector*; DAFF: Pretoria, South Africa, 2018. Available online: https://www.daff.gov.za/Daffweb3/Portals/0/StatisticsandEconomicAnalysis/StatisticalInformation/TrendsintheAgriculturalSector2017.pdf (accessed on 28 June 2019).

29. Meissner, H.H.; Scholtz, M.M.; Engelbrecht, F.A. Sustainability of the South African livestock sector towards 2050 part 2: Challenges, changes and required implementations. *S. Afr. J. Anim. Sci.* **2013**, *43*, 298–319. [CrossRef]

30. StatsSA. *Community Survey 2016 Agricultural Households*; StatsSA: Pretoria, South Africa, 2016. Available online: http://www.statssa.gov.za/?page_id=964 (accessed on 28 June 2019).

31. Bo, S. A literature survey on environmental Kuznets curve. *Energy Procedia* **2011**, *5*, 1322–1325. [CrossRef]

32. Kijima, M.; Nishide, K.; Ohyama, A. Economic models for the environmental Kuznets curve: A survey. *J. Econ. Dyn. Control* **2010**, *34*, 1187–1201. [CrossRef]

33. Shahbaz, M.; Kumar Tiwari, A.; Nasir, M. The effects of financial development, economic growth, coal consumption and trade openness on CO_2 emissions in South Africa. *Energy Policy* **2013**, *61*, 1452–1459. [CrossRef]

34. Inglesi-Lotz, R.; Bohlmann, J. Environmental Kuznets curve in South Africa: To confirm or not to confirm? In Proceedings of the Ecomod Conference, Bali, Indonesia, Bali, Indonesia, 6–18 July 2014; pp. 1–17.

35. Kaika, D.; Zervas, E. The Environmental Kuznets Curve (EKC) theory. Part B: Critical issues. *Energy Policy* **2013**, *62*, 1403–1411. [CrossRef]

36. Lin, B.; Omoju, O.E.; Nwakeze, N.M.; Okonkwo, J.U.; Megbowon, E.T. Is the environmental Kuznets curve hypothesis a sound basis for environmental policy in Africa ? *J. Clean. Prod.* **2016**, *133*, 712–724. [CrossRef]

37. Parker, H. *Understanding Patterns of Climate Resilient Economic Development in Rwanda*; ODI Annual Report: London, UK, 2015.

38. Bouvier, R.A. *Air Pollution and Per Capita Income: A Disaggregation of the Effects of Scale, Sectoral Composition, and Technological Change*; Report No.: Working Paper, Series; Brock, W.A., Ed.; University of Massachusetts at Amherst: Boston, MA, USA, 2004; p. 84.

39. Gill, A.R.; Viswanathan, K.K.; Hassan, S. The Environmental Kuznets Curve (EKC) and the environmental problem of the day. *Renew. Sustain. Energy Rev.* **2018**, *81*, 1636–1642. [CrossRef]

40. Baek, J. Environmental Kuznets curve for CO_2 emissions: The case of Arctic countries. *Energy Econ.* **2015**, *50*, 13–17. [CrossRef]

41. Kaika, D.; Zervas, E. The Environmental Kuznets Curve (EKC) theory-Part A: Concept, causes and the CO_2 emissions case. *Energy Policy* **2013**, *62*, 1392–1402. [CrossRef]

42. Balaguer, J.; Cantavella, M. Estimating the environmental Kuznets curve for Spain by considering fuel oil prices (1874–2011). *Ecol. Indic.* **2016**, *60*, 853–859. [CrossRef]
43. Alam, M.M.; Murad, M.W.; Noman, A.H.; Ozturk, I. Relationships among carbon emissions, economic growth, energy consumption and population growth: Testing Environmental Kuznets Curve hypothesis for Brazil, China, India and Indonesia. *Ecol. Indic.* **2016**, *70*, 466–479. [CrossRef]
44. Apergis, N. Environmental Kuznets curves: New evidence on both panel and country-level CO_2 emissions. *Energy Econ.* **2016**, *54*, 263–271. [CrossRef]
45. Al-Mulali, U.; Ozturk, I. The investigation of environmental Kuznets curve hypothesis in the advanced economies: The role of energy prices. *Renew. Sustain. Energy Rev.* **2016**, *54*, 1622–1631. [CrossRef]
46. Ahmad, N.; Du, L.; Lu, J.; Wang, J.; Li, H.Z.; Hashmi, M.Z. Modelling the CO_2 emissions and economic growth in Croatia: Is there any environmental Kuznets curve? *Energy* **2017**, *123*, 164–172. [CrossRef]
47. Özokcu, S.; Özdemir, Ö. Economic growth, energy, and environmental Kuznets curve. *Renew. Sustain. Energy Rev.* **2017**, *72*, 639–647. [CrossRef]
48. Churchill, S.A.; Inekwe, J.; Ivanovski, K.; Smyth, R. The Environmental Kuznets Curve in the OECD: 1870–2014. *Energy Econ.* **2018**, *75*, 389–399. [CrossRef]
49. World Bank. World Development Indicators: Agriculture, Forstery and Fishing (% added of GDP)—South Africa. 2019. Available online: https://data.worldbank.org/indicator/NV.AGR.TOTL.ZS?locations=ZA&view=chart (accessed on 3 July 2019).
50. He, J.; Richard, P. Environmental Kuznets curve for CO_2 in Canada. *Ecol Econ.* **2010**, *69*, 1083–1093. [CrossRef]
51. Pesaran, M.H.; Pesaran, B. *Working with Microfit 4.0: Interactive Econometric Analysis*; Oxford University Press: Oxford, UK, 1997.
52. Pesaran, M.H.; Shin, Y.; Smith, R.J. Bounds testing approaches to the analysisof level relationships. *J. Appl. Econom.* **2001**, *16*, 289–326. [CrossRef]
53. Pesaran, M.H.; Shin, Y. An autoregressive distributed lag modelling approachto cointegration analysis. In *Econometrics and Economic Theoryin 20th Century: The Ragnar Frisch Centennial Symposium*; Strom, S., Ed.; Cambridge University Press: Cambridge, UK, 1997.
54. Said, S.E.; Dickey, D.A. Testing for unit roots in autoregressive-moving averagemodels of unknown order. *Biometrika* **1984**, *71*, 599–607. [CrossRef]
55. Lütkepohl, H. *Structural Vector Autoregressive Analysis for Cointegrated Vari-Ables*; Springer: Heidelberg/Berlin, Germany, 2006; pp. 73–86.
56. Shahbaz, M.; Aviral, T.; Nasir, M. *The Effects of Financial Development, Economic Growth, Coal Consumption and Trade Openness on Environment Performance in South Africa*; Federal Bureau of Statistics, Government of Pakistan: Islamabad, Pakistan, 2011.
57. Nasr, B.A.; Gupta, R.; Sato, R.J. Is there an Environmental Kuznets Curve for South Africa? A co-summability approach using a century of data. *Energy Econ.* **2015**, *52*, 136–141. [CrossRef]
58. Shahbaz, M.; Adebola, S.; Ozturk, I. Environmental Kuznets Curve hypothesis and the role of globalization in selected African countries. *Ecol. Indic.* **2016**, *67*, 623–636. [CrossRef]

Review

The Future of Semi-Arid Regions:
A Weak Fabric Unravels

Robert J Scholes

Global Change Institute, University of the Witwatersrand, Johannesburg 2050, South Africa;
bob.scholes@wits.ac.za

Received: 14 February 2020; Accepted: 12 March 2020; Published: 13 March 2020

Abstract: The regions of the world where average precipitation is between one fifth and half of the potential plant water demand are termed 'semi-arid'. They make up 15.2% of the global land surface, and the approximately 1.1 billion people who live there are among the world's poorest. The inter-annual variability of rainfall in semi-arid regions is exceptionally high, due to intrinsic features of the global atmospheric circulation. The observed and projected climate trends for most semi-arid regions indicate warming at rates above the global mean rate over land, increasing evaporative demand, and reduced and more variable rainfall. Historically, the ecosystems and people coped with the challenges of semi-arid climates using a range of strategies that are now less viable. Semi-arid ecosystems are by definition water limited, generally only suitable for extensive pastoralism and opportunistic cropping, unless irrigation supplementation is available. The characteristics of dryland plant production in semi-arid ecosystems, as they interact with climate change and human systems, provide a conceptual framework for why land degradation is so conspicuous in semi-arid regions. The coupled social-ecological failures are contagious, both within the landscape and at regional and global scales. Thus, semi-arid lands are a likely flashpoint for Earth system changes in the 21st century.

Keywords: vulnerability; contagion; climate change; land degradation

1. Introduction: Regions with Semi-Arid Climates as a Distinct Global Entity

The Food and Agriculture Organisation Agro-Ecological Zone concept defines those parts of the world where the annual precipitation (P) sums to between one fifth and one half of the potential evapotranspiration (E_p) as 'semi-arid' (i.e., $0.2 < P/E_p \leq 0.5$) [1]. This translates, on average, to between 60 and 180 days of plant growth opportunity. At the lower end this is just enough for some fast-growing annual crops. At the upper end it provides a reasonably reliable annual crop yield in most years.

This paper focuses on the 'hot semi-arid regions' (Mean Annual Temperature > 18 °C, which serves as a separator between the neotropical drylands and the temperate to sub-arctic drylands). These areas, corresponding to the Köppen class BSh, are clustered around the tropics of Cancer and Capricorn, where the descending limbs of the Hadley cells, to the north and south of the equator, result in extensive dry areas on all continents. In addition, there are cold semi-arid areas at higher latitudes (Köppen class BSk). These face a somewhat different set of climatic and social futures and are thus not substantially treated here (Figure 1).

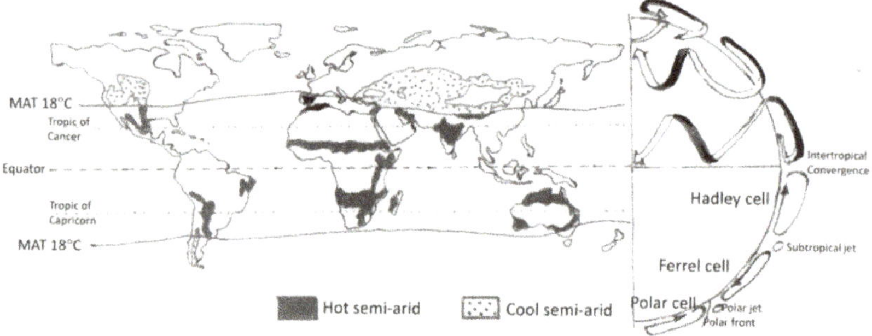

Figure 1. The distribution of semi-arid areas in relationship to the major features of the global atmospheric circulation. The schematic to the right shows in the southern hemisphere a stylised cross section of the globe and the tropospheric cells. The northern hemisphere has a representation of the three-dimensional nature of the circulation, with the surface winds in white and the upper-atmosphere winds in black. The hot semi-arid areas have a mean annual temperature (MAT) above 18 °C and are the main focus of this paper. Cool semi-arid areas share some of their water-limited characteristics, but in addition have a strong control of primary production by low temperatures in winter.

The hot semi-arid regions have characteristic unifying climatic, ecological and social-system features that mark them out as a coherent global category with respect to the challenges of the Anthropocene, including but not restricted to climate change. The key climate characteristics are high solar radiation and thus high mean daytime air temperatures, due to the low cloud cover and subtropical location [2]; overall low atmospheric humidity and rainfall, because of the predominance of high-pressure systems (descending air masses) [3]; prominent within-year rainfall seasonality, due to the north-south migration of the intertropical convergence zone, typically consisting of a hot 'summer' wet season and a warm 'winter' dry season (though winter-wet variants also occur, as do 'monsoonal' systems with two wet seasons a year, one at each equinox) [4]; and high inter-annual variability of rainfall, often linked to variable modes of the global climate system, such as the ENSO phenomenon [5]. The coefficient of variation of inter-annual rainfall is often 30% or higher—a threshold which ecologists have used to differentiate 'equilibrium systems' from 'non-equilibrium systems' [6,7].

The ecological landscape in semi-arid regions is often notably spatially patchy [8]. This patchiness has many sources, operating at a variety of scales. Precipitation in semi-arid lands occurs mostly as discrete convective storms, which have a characteristic spatial footprint of several kilometres [9]. A second major source of spatial variation is geomorphology (the landform and soil). The prominence of geomorphology in semi-arid lands is ultimately due to the absence of glaciation and intermediate pace of pedogenesis during the Pleistocene. Wetter warm landscapes (i.e., the moist tropics) tend to be flatter and dominated by homogeneously highly weathered, nutrient-deleted soils, because weathering and pedogenesis has run its course. Drier, hotter landscapes (i.e., arid lands and hyperarid deserts) show little soil development and differentiation. In between, semi-arid landscapes typically have a distinct drainage pattern, with soils arranged in a topographical sequence from ridge crest to valley floor, at a scale of a few kilometres (the original reference to the 'catena concept' is [10]; there has been much work subsequently). There are several soil-animal-vegetation feedback processes which act to reinforce this basic 'climate by geomorphology' spatial template [11]. The interaction between the temporal variability of the climate with the spatial variability of the landscape, is a central feature of the natural and human ecology of semi-arid regions [12].

The warm semi-arid regions of the world are home to about 1.1 billion people, many of whom are rural dwellers. With some notable exceptions (e.g., Australia and the southern United States of America), the median household income for semi-arid areas is lower than that of sub-humid areas [13]. Since crop agriculture without irrigation is a risky enterprise in semi-arid lands, the main agricultural land use is pastoralism, involving cattle, sheep, goats, and camels. Opportunistic dryland cropping is widespread, especially towards the wetter end of the spectrum and where there are institutions that allow land uses to persist despite seasonal crop failures. A consequence of the mismatch between the high natural temporal variability and the often slower pace of adaptive response by human systems has been episodes of extensive land degradation, resulting from the over-optimistic expansion of crops or herds during period of rainfall abundance, followed by their collapse in times of multi-year drought. Well-known examples, among many, include the dustbowl of the southern USA, the loess plateau of China, the mallee lands of south-east Australia, and the little Karoo in South Africa [14–17].

A frequently reported feature of semi-arid lands is economic and political marginalisation. The economic yield per unit area of land is low, thus the population density is low. Semi-arid lands are frequently distant from the main population centres where policy is made. While the people of semi-arid lands often have a strong identity and sense of place, their influence on larger-scale political processes is usually weak [18–20].

2. The Climatic Determinants of Semi-Aridity and How They are Changing

The Hadley cells are atmospheric circulation structures on either side of the equator (strictly speaking, the energy equator, the line where extra-terrestrial solar input is maximum, rather than the line of 0° latitude; the energy equator moves north and south with the seasons). When moist equatorial air is warmed and rises, the atmospheric profile becomes unstable, producing clouds and the rainfall that supports the equatorial forests. The now-drier air moves poleward at the tropopause, under the influence of the equator-pole temperature gradient. Several thousands of kilometres further north or south it descends, forming a broad band of discrete high-pressure cells that are relatively dry. Because of the low cloud cover fraction and location are relatively near to the tropics, these lands tend to be hot. The air mass then circulates back towards the equator, closer to the land surface, gathering moisture as it goes. The entire paired structure of the northern and southern Hadley circulation moves northward in the boreal summer and southward in the austral summer, in response to the apparent movement of the latitudinal position of the sun, but with a delay and an attenuated amplitude. The band where which the northern and southern Hadley cells abut is called the inter-tropical convergence zone (ITCZ) and is associated with torrential rains. The ITCZ is a seasonally and geographically meandering line, bending poleward over land and equatorward over oceans, not perfectly symmetrical between the northern and southern hemispheres, and behaving slightly differently from year to year, depending on global and regional patterns of oceanic temperature. The position of the ITCZ is a key determinant of rainfall in the semi-arid regions, since they lie between its poleward limit and the poleward limit of its antithesis—the divergence zone between the Hadley and Ferrell cells, which are the next generalised circulation structure toward the poles. The dynamics of the Hadley cells explain both the strong seasonality and the high inter-annual variability in semi-arid regions [21,22].

The Hadley circulation is a prominent and consistent feature of the global climate system but is not invariant. In simplified mathematical models of the global circulation it emerges in response to the establishment of an equator-pole temperature gradient, and its breadth and strength is influenced by that gradient [23]. In the more complex, three-dimensional simulations of 21st century climates, precipitation generally increases globally in response to rising greenhouse gas concentrations and thus rising global mean air temperature, but in general the rain falling in semi-arid areas, i.e., those seasonally covered by the descending limbs of the Hadley cells, decreases [24,25]. At the same time in these zones, air temperature rises, wind-fields strengthen, humidity decreases and net solar radiation at the Earth surface may increase—all of which cause the potential evaporation to increase. Thus, water balance indicators in semi-arid lands, such as soil moisture duration and the P/E_p ratio decrease in future, with high consistency across many model platforms and scenarios. All models agree that the variability structure of individual rainfall events also changes, most indicating fewer rainfall events, but a relatively greater fraction of rain falling in the form of intense events. Most models also project the inter-annual variability of rainfall in semi-arid lands to increase, as a result of re-organisations in the coupled ocean-atmosphere circulation, leading to greater persistence of modes associated with either extended droughts or high-rainfall periods, but there is substantial disagreement between models regarding the details [26].

This review is less focussed on the idea that the semi-arid lands are expanding ('desert encroachment') than on the intensification of biophysical and social processes within the historical footprint of the hot semi-arid regions which increase the risk of social-ecological systems failure.

3. Plant Production Dynamics in Semi-Arid Ecosystems

In warm semi-arid lands, where E_p is consistently high, the relationship between net plant production (NPP) and growing season rainfall is very nearly linear and has a similar slope in all parts of the world, as is theoretically expected given the prominence of water limitation in this environment [27–29]. Since human welfare in semi-arid regions ultimately depends to a large degree on plant production, the directness of the climate-production relationship means high confidence in projecting future outcomes—but the apparent simplicity of the annual-scale predictive model given by (NPP = $a^*(P/E_p)$ + b) belies some underlying subtleties.

A useful way of thinking about the relation is that the slope (a) is a measure of plant water use efficiency (Many researchers define 'rain use efficiency' as RUE = NPP/P. This is problematic, given that the linear relation seldom passes exactly through the origin (i.e., b ≠ 0). As a result, RUE thus defined apparently changes as function of P. This is just a mathematical truism, with no inherent ecological insight) while the intercept (b) is a measure of the landscape water partitioning, either moisture carried over between years, or lost to the landscape unproductively as runoff or evaporation from the soil surface. To see this, express the intercept in terms of the x-axis rather than the y-axis intercept, NPP = $a^*(P/E_p - c)$, where c is −a/b. Water use efficiency at the plant scale is sensitive to plant functional type—for instance, C4 grasses have a higher value than C3 trees—but also to nutrient supply. If nutrients are sufficient, the plants can grow at their maximum rate; if not, the rate is reduced. The intercept parameter is sensitive to soil water holding capacity; it is higher on deep, sandy soils and lower on shallow or clayey soils [30].

The pragmatic version of the above relationship (Given that across many semi-arid regions E_p is quite consistently around 1500 mm·y^{-1}, and in semi-arid rangelands AGNPP is about half of NPP, approximate conversions can be made between the various forms of the basic linear relationship between production and water availability. Furthermore, gross primary production (GPP) is about twice NPP, and since GPP = $\varepsilon^*\sum$(FAPAR*PAR), a logical bridge can be built to satellite-derived estimates of plant productivity based on seasonally accumulated greenness indices, such as FAPAR or NDVI) uses aboveground NPP (AGNPP) and mean annual precipitation (MAP) as proxies (i.e., AGNPP = a^*MAP + b). For semi-arid grassland regions around the world, the relationship is essentially identical when determined over a number of sites, each measured for several years, and spanning a MAP gradient. These 'regional' relations have a value of the slope constant 'a' if around 0.65 g·m^{-2} mm^{-1} and the intercept 'b' around −50 g·m^{-2}. The regional relations explain a high fraction of observed variance (typically R^2 > 0.8). However, when production is measured at a single site over many years, and annual production in each year is related to the rainfall for the year at that site, the correlation is much weaker and the parameters differ systematically from those derived for the across-site case: the slopes (a) are lower and the incept (b) is higher [29]. Thus, the regional mean relation cannot be reliably applied to individual sites. The divergence between the local (temporal) and regional (spatial) relationship is relevant to the issue of semi-arid landscape vulnerability and has been explained as follows. Over a long period of time (decades or centuries), the nutrient stock at each location comes into equilibrium with the NPP which can be supported by the long-term water balance at that site. Thus, the regional relation has a tight linear fit, quite invariant across very different regions. In any given year at a given site, however, the nutrient supply is either less than needed (in a high rainfall year) or more than needed (a low rainfall year), thus the slope flattens. There are also carry-over effects at a site between years, which have the consequence of raising the intercept. These dynamics mean that it is possible for the site equilibrium to lag behind climate trends, presenting as land degradation [29,31].

The contemporary practice is to define land degradation as a persistent decrease in the ecosystem services ('the benefits that humans derive from nature') delivered by the land, where persistence means a timescale considerably longer than either the natural variation or the coping capacity of human institutions [32,33]. Many ecosystem services depend directly or indirectly on plant production, biomass, and cover—including, to a substantial degree, those benefits based on biodiversity. Therefore, degradation requires either that the expected value of P/E_p decreases (a robust prediction for warm semi-arid lands in the 21st century), or that the *parameters* a or b change in a persistent way. If a is a measure of vegetation functional type composition and/or nutrient availability as suggested above, degradation can result either from a composition shift to less water-efficient types, or a reduction of soil nutrient supplying capacity. The intercept (b) can be reduced by soil loss, soil surface sealing, or a change in the proportion of water evaporating from the soil rather than transpiring from plants. All these mechanisms of persistent parameter change are increasingly likely to occur when the rainfall, over a multi-year period, is below expectation, and is more intense when it occurs. The relationship between drought, land use, and land degradation is complex, but all authorities agree that such a linkage exists and is important [34,35]. This means that increases in inter-annual rainfall variability and decreases in P/E_p, widely predicted outcomes in semi-arid lands in this century, are likely also to result in further degradation, particularly if the human system fails to adapt to low rainfall periods rapidly and appropriately.

4. Plant and Animal and Coping Strategies and Their Limits

The plants and animals in semi-arid lands have evolved under conditions of water scarcity, and therefore have adapted to them. Semi-arid systems are helpfully thought of as being pulsed by water availability—there are more-or-less discrete periods of water availability, of variable duration, interspersed with periods of non-availability, during which physiological activity is greatly constrained [36]. Some of this temporal variability is predictable, such as between rainy and dry seasons within the year. Some is less predictable, such as the variation between years, or the occurrence of dry spells within the supposed wet season. A wide range of life history strategies have evolved in semi-arid lands to cope with this variation in water availability. Some plants (annual ephemerals) avoid drought by having desiccation-resistant seeds, and germinating, growing and reproducing rapidly in response to wetting events. Others have a perennial strategy but die back or shed leaves in the dry season to restrict water loss. Some store water in stems and bulbs. Some tolerate extreme desiccation, or eke out the water supply that they have, or tap into groundwater (for a recent review, see [37]). Some keep the leaf water potential from falling by closing their stomata (isohydric) while others prioritise production and keep their stomata open while allowing leaf water potential to fall (anisohydric) [38]. All these strategies are demonstrably viable, but each only within the historically experienced range of the multi-dimensional temporal water availability regime. A change in the regime, or a change in other evolutionary pressures (such as fire or herbivory) leads to a change in the functional type composition, and in the medium term often to a loss in biodiversity. Conversely, the maintenance of functional diversity confers some insurance against water supply variability [39,40]. Woody-plant mortality has been observed following atypical drought in dry forests, and regeneration may be impeded [41].

Animals in semi-arid landscapes also exhibit a range of physiological and behavioural mechanisms to cope with the scarcity, variability, and unreliability of water supply. Particularly important is the capacity to relocate to places where water is available. The restriction of this capability by fragmentation of the habitat or otherwise impeding animal movement is a key contributor to a reduction in arid land resilience [42,43]. Protracted periods of extremely high temperatures are a serious threat to both the survival and productivity of warm-blooded animals, including humans and their livestock, in hot semi-arid areas [44].

5. The Failure of Coupled Socio-Ecological Systems in Semiarid Regions

The IPBES Land Degradation and Restoration Assessment demonstrated that examples of land degradation and attendant human wellbeing loss can be found in every terrestrial system, worldwide—but that some ecosystems, notably including the semi-arid landscapes, are more vulnerable than others [33].

Partly this can be explained in biophysical terms, as the presence in semi-arid ecosystems of several processes where the state of the system can easily stray across a degradation threshold, beyond which recovery is slow. It can also be partly explained in terms of the poverty and political marginalisation of many social systems in arid lands. However, a full understanding of the vulnerability of semiarid lands also needs to consider the interaction between the social and ecological systems. It can be argued that when human and ecological systems have coexisted for many generations without significant external intervention, they should have come into a sustainable equilibrium, and there are cases which support that contention [45]. However, there have been major changes in semi-arid land use worldwide over the past two centuries, which mean that contemporary social-ecological systems may not yet be fully co-adapted, even if they once were. Furthermore, few arid lands are now governed entirely by their internal dynamics. Global climate change is only one intrusive external factor—others include global trade, the demand for mineral resources, and the upscaling and centralisation of governance.

Three technologies have had substantial impact in semiarid lands. The first was the development of fencing, enabling the restriction of animal movement and the privatisation of the commons. The second was the ability to tap groundwater using boreholes and pumps. The third was the introduction of highly bred domesticated herbivores, usually non-native, and the consequent displacement of indigenous herbivores. All three, along with the political desire to settle and control nomadic pastoralists, have contributed to the reduction of the spatial adaptability of semi-arid lands, and thus a loss of resilience. The increase in rigidity has a temporal element as well. Infrastructural investment, loans, drought assistance, and private property ownership rules can all act to allow a land use to persist at an intensity mismatched to a transient change in ecological capacity, resulting in a persistent reduction in that capacity [33].

The factors described above and listed below combine and interact to make semi-arid lands exceptionally vulnerable to social and ecological failure: their political and economic marginalisation; the relative slowness in many to undergo the demographic transition to a lower birth rate matching the modern reduced death rate that has been observed in other lands, worldwide; the concurrence of climate change stresses; inherent ecological features; and maladapted social systems. A potential consequence, already demonstrable in some accounts, is human conflict and migration [46,47]. The evidence that this is an inevitable consequence of land degradation is extremely mixed [48].

Identifying contagious processes (i.e., sequential degradation and reduction in human wellbeing in adjacent or otherwise spatially connected areas) is a key indicator for the detection of land degradation. Early detection is an important avoidance criterion, since prevention is inevitably cheaper and more effective than remediation after degradation has become severe [33]. At a local scale, it is observed in high-resolution satellite images that degradation patches are coherent at spatial sales of up to several kilometres, rather than involving random pixels. This suggests a contagious mechanism is at work. For instance, unless the stocking rate is decreased, reductions in NPP at a pixel scale increases the grazing pressure on adjacent patches. Contagion mechanisms also occur at larger scales—for instance, the displacement of people is one such mechanism. Another long-distance teleconnection involves the airborne transport of dust derived from degraded semiarid lands [49,50]. A third involves the silt load and flooding of rivers [51]. There are several hypothetical mechanisms by which land degradation in semi-arid lands could have regional-scale climatic feedback consequences, especially rainfall reduction [52,53]. These hypotheses are neither fully tested and established, nor unequivocally disproved.

6. Interventions in Semi-Arid Lands

6.1. Actions That are Unlikely to Help

Massive tree-planting schemes have been widely advocated in semi-arid lands, often as cheap, no-pain climate mitigation schemes, thinly disguised as human-wellbeing enhancements or land restoration interventions [54]. Semi-arid landscapes have an inherent upper limit to the tree cover they can support [55]; thus, trying to establish extensive closed-canopy forests in semi-arid areas will not succeed. Even increasing the tree cover to the limit which can be supported comes at the expense of forage production, crop production and water yield—this is unsurprising, given that water limitation is the unifying feature of such landscapes. If afforestation is conducted using monocultures, typically of exotic species, it also imposes a biodiversity penalty [56]. Apart from the likelihood of expensive, delaying, and damaging afforestation failures, replacing a typically bright land surface with a dark perennial tree-covered surface, in an area of low productivity but high solar radiation, almost certainly leads to more, rather than less, global warming.

There is a great temptation, in times of water scarcity in semi-arid lands, to supply groundwater through boreholes. This has two limitations. First, average groundwater recharge in semi-arid lands typically amounts to only a few percent of MAP [57]. This sets a limit on the long-term sustainability of most such schemes. Secondly, perennial water supplies act as a focus of animal and human concentration, potentially leading to local degradation [58].

The perceived negative welfare consequences of urbanisation result in calls to promote rural development, in order that the population stay on the land. There are very few examples, worldwide, where such a strategy has succeeded in the long term. The move to cities is a rational choice in economies where land-based activities (i.e., agriculture) are a shrinking part of the mix. In many instances the rural landscape is 'full' in terms of the availability of new land to exploit, and subdivision of the land to accommodate rural population growth is counterproductive given the low per area productivity. Despite the problems associated with rapid urbanisation, the movement to towns helps to relieve pressure on the land and may be a necessary response to a declining rural resource capacity. Notwithstanding the high cost of developing urban infrastructure, services are usually more cost-efficiently provided in more densely settled areas than in sparsely settled rural contexts.

6.2. Interventions with Greater Promise of Success

Land use systems that either replicate the adaptive capacity of the natural or traditional social-ecological systems they replace, or simulate it other ways, can rebuild and enhance the resilience of semi-arid lands. For instance, rotational grazing schemes offer restoration advantages over fixed, sedentary grazing [59], controllable water-points are better for avoiding degradation than permanent water provision [60], and schemes for rapid, low-cost stocking rate adjustment or stock movement are better than the import of supplemental feed that allows the overgrazing of natural rangelands [61,62]. Examples of artificially introduced flexibility include meta-population management of wildlife in place of natural migration [63], and a combination of drought early warnings and cropland set-aside schemes that incentivise reduction in the cultivated area in predicted low rainfall years, rather than yield insurance schemes that incentivise speculative planting [64]. Several 'conservation agriculture' techniques, including low or no-till cropping, allow crops to make better use of a deteriorating water balance, as does shifting to more drought-and-heat resistant varieties and crop species [64].

Semi-arid landscapes in many parts of the world are undergoing spontaneous or deliberate rewilding. For instance, the outback of Australia is depopulating as livestock-based farm enterprises become less labour intensive, and sometimes not economically viable [65]. Southern Africa has witnessed a dramatic shift from cattle and sheep pastoralism to mixed indigenous herbivores, servicing the wildlife-based tourism and hunting industry [66]. In an increasingly urbanised world, relatively natural and sparsely populated spaces are a scarce and valuable commodity. Especially if rewilding is accompanied by larger spatial scales of management, such land uses are likely to be more able to tolerate lower and more variable rainfall, higher temperatures and multi-year droughts than land uses based on domestic livestock or crop production alone.

Many semi-arid landscapes worldwide are seeing a shift from an economy based on plant and animal production to one based on energy generation. Semi-arid lands are ideal for the capture of solar energy. The near-equatorial location of hot semi-arid areas results in high extra-terrestrial solar fluxes, and dryness in both hot and cold semi-arid areas means low cloud cover. In contrast to more arid regions, they are typically closer to energy users and have some population, and thus have a basic infrastructure and local energy demand. Land in semi-arid landscapes is relatively cheap. Some semi-arid lands also have wind energy potential, with fewer of the visual and noise complications of wind power networks in more densely populated landscapes. The combination of wind and solar power helps to cover for some of the intermittency problems of either implemented alone. Semi-arid lands have been mooted as locations for the production of bioenergy or the sequestration of carbon. This is not out of the question, depending on the energy or carbon price, but is constrained by the inherently low NPP, its inter-annual variability, and the low carbon storage potential in hot soils. These result in a low carbon density for sequestration, and a low energy density for biofuels.

Funding: This research received no external funding.

Acknowledgments: Travel of R.J.S. to the conference at College Station 26–27 October 2018, was supported by Texas A & M University.

Conflicts of Interest: There are no conflicts of interest to report.

References

1. Higgins, G.M.; Kassam, A.H. The FAO agro-ecological zone approach to determination of land potential. *Pedologie* **1981**, *31*, 147–168.
2. Black, J.N. The distribution of solar radiation over the earth's surface. *Arch. Meteorol. Geophys. Bioklimatol. Ser. B* **1956**, *7*, 165–189. [CrossRef]
3. Dima, I.M.; Wallace, J.M. On the seasonality of the Hadley cell. *J. Atmos. Sci.* **2003**, *60*, 1522–1527. [CrossRef]
4. Walsh, R.P.D.; Lawler, D.M. Rainfall seasonality: Description, spatial patterns and change through time. *Weather* **1981**, *36*, 201–208. [CrossRef]
5. Camberlin, P.; Janicot, S.; Poccard, I. Seasonality and atmospheric dynamics of the teleconnection between African rainfall and tropical sea-surface temperature: Atlantic vs. ENSO. *Int. J. Climatol. A J. R. Meteorol. Soc.* **2001**, *21*, 973–1005. [CrossRef]
6. Ellis, J.E.; Swift, D.M. Stability of African pastoral ecosystems: Alternate paradigms and implications for development. *J. Rangel. Manag.* **1988**, *41*, 450–459. [CrossRef]
7. Von Wehrden, H.; Hanspach, J.; Kaczensky, P.; Fischer, J.; Wesche, K. Global assessment of the non-equilibrium concept in rangelands. *Ecol. Appl.* **2012**, *22*, 393–399. [CrossRef] [PubMed]
8. Aguiar, M.R.; Sala, O.E. Patch structure, dynamics and implications for the functioning of arid ecosystems. *Trends Ecol. Evol.* **1999**, *14*, 273–277. [CrossRef]
9. Taylor, C.M.; Birch, C.E.; Parker, D.J.; Dixon, N.; Guichard, F.; Nikulin, G.; Lister, G.M. Modeling soil moisture-precipitation feedback in the Sahel: Importance of spatial scale versus convective parameterization. *Geophys. Res. Lett.* **2013**, *40*, 6213–6218. [CrossRef]
10. Milne, G.A. A provisional soil map of East Africa. *East Afr. Agric. Res. Stn.* **1936**, *88*, 465.
11. Scholes, R.J.; Bond, W.J.; Eckhardt, H.C. Vegetation dynamics in the Kruger ecosystem. In *The Kruger Experience: Ecology and Management of Savanna Heterogeneity*; Island Press: Washington, DC, USA, 2003; pp. 242–262.
12. Mueller, E.N.; Wainwright, J.; Parsons, A.J. Spatial variability of soil and nutrient characteristics of semi-arid grasslands and shrublands, Jornada Basin, New Mexico. *Ecohydrol. Ecosyst. Land Water Process Interact. Ecohydrogeomorphol.* **2008**, *1*, 3–12. [CrossRef]
13. Hallegatte, S.; Rozenberg, J. Climate change through a poverty lens. *Nat. Clim. Chang.* **2017**, *7*, 250–256. [CrossRef]

14. Cook, B.I.; Miller, R.L.; Seager, R. Amplification of the North American "Dust Bowl" drought through human-induced land degradation. *Proc. Natl. Acad. Sci. USA* **2009**, *106*, 4997–5001. [CrossRef] [PubMed]

15. Fu, B.; Wang, S.; Liu, Y.; Liu, J.; Liang, W.; Miao, C. Hydrogeomorphic ecosystem responses to natural and anthropogenic changes in the Loess Plateau of China. *Annu. Rev. Earth Planet. Sci.* **2017**, *45*, 223–243. [CrossRef]

16. Chartres, C. *Australia's Land Resources at Risk*; Cambridge University Press: Cambridge, UK, 1987; pp. 7–26.

17. Hoffmann, T.; Todd, S.; Ntshona, Z.; Turner, S. *Land Degradation in South Africa*; University of Cape Town: Cape Town, South Africa, 2014.

18. Nel, E.; Hill, T. Marginalisation and demographic change in the semi-arid Karoo, South Africa. *J. Arid Environ.* **2008**, *72*, 2264–2274. [CrossRef]

19. Pavanello, S. *Pastoralists' Vulnerability in the Horn of Africa: Exploring Political Marginalisation, Donors' Policies and Cross-Border Issues–Literature Review*; Humanitarian Policy Group (HPG) Overseas Development Institute: London, UK, 2009.

20. Blaikie, P. *The Political Economy of Soil Erosion in Developing Countries*; Routledge: London, UK, 2016.

21. Schneider, T.; Bischoff, T.; Haug, G.H. Migrations and dynamics of the Intertropical Convergence Zone. *Nature* **2014**, *513*, 45–53. [CrossRef]

22. Nicholson, S.E. The ITCZ and the Seasonal Cycle over Equatorial Africa. *Bull. Am. Meteorol. Soc.* **2018**, *99*, 337–348. [CrossRef]

23. Gill, A.E. Some simple solutions for heat-induced tropical circulation. *Q. J. R. Meteorol. Soc.* **1980**, *106*, 447–462. [CrossRef]

24. Intergovernmental Panel on Climate Chang. *The Physical Science Basis-Summary for Policymakers, Observed Changes in the Climate System*; IPCC: Geneva, Switzerland, 2013; p. 15.

25. Fernandez, J.P.R.; Franchito, S.H.; Rao, V.B. Future changes in the aridity of South America from regional climate model projections. *Pure Appl. Geophys.* **2019**, *176*, 2719–2728. [CrossRef]

26. Chen, C.; Cane, M.A.; Wittenberg, A.T.; Chen, D. ENSO in the CMIP5 simulations: Life cycles, diversity, and responses to climate change. *J. Clim.* **2017**, *30*, 775–801. [CrossRef]

27. McNaughton, S.J. Ecology of a grazing ecosystem: The Serengeti. *Ecol. Monogr.* **1985**, *55*, 259–294. [CrossRef]

28. Sala, O.E.; Parton, W.J.; Joyce, L.A.; Lauenroth, W.K. Primary production of the central grassland region of the United States. *Ecology* **1988**, *69*, 40–45. [CrossRef]

29. Sala, O.E.; Gherardi, L.A.; Reichmann, L.; Jobbagy, E.; Peters, D. Legacies of precipitation fluctuations on primary production: Theory and data synthesis. *Philos. Trans. R. Soc. B Biol. Sci.* **2012**, *367*, 3135–3144. [CrossRef] [PubMed]

30. Scholes, R.J. Convex relationships in ecosystems containing mixtures of trees and grass. *Environ. Resour. Econ.* **2003**, *26*, 559–574. [CrossRef]

31. Van den Hoof, C.; Verstraete, M.; Scholes, R.J. Differing Responses to Rainfall Suggest More Than One Functional Type of Grassland in South Africa. *Remote Sens.* **2018**, *10*, 2055. [CrossRef]

32. Scholes, R.J. Syndromes of dryland degradation in southern Africa. *Afr. J. Range Forage Sci.* **2009**, *26*, 113–125. [CrossRef]

33. Montanarella, L.; Scholes, R.; Brainich, A. (Eds.) *The IPBES Assessment Report on Land Degradation and Restoration*; Secretariat of the Intergovernmental Science-Policy Platform on Biodiversity and Ecosystem Services: Bonn, Germany, 2018.

34. Vicente-Serrano, S.M.; Cabello, D.; Tomás-Burguera, M.; Martín-Hernández, N.; Beguería, S.; Azorin-Molina, C.; Kenawy, A.E. Drought variability and land degradation in semiarid regions: Assessment using remote sensing data and drought indices (1982–2011). *Remote Sens.* **2015**, *7*, 4391–4423. [CrossRef]

35. Herrmann, S.M.; Hutchinson, C.F. The scientific basis: Links between land degradation, drought and desertification. In *Governing Global Desertification*; Routledge: London, UK, 2016; pp. 31–46.

36. Williams, C.A.; Hanan, N.; Scholes, R.J.; Kutsch, W. Complexity in water and carbon dioxide fluxes following rain pulses in an African savanna. *Oecologia* **2009**, *161*, 469–480. [CrossRef]

37. Osakabe, Y.; Osakabe, K.; Shinozaki, K.; Tran, L.S.P. Response of plants to water stress. *Front. Plant Sci.* **2014**, *5*, 86. [CrossRef]

38. Mirfenderesgi, G.; Matheny, A.M.; Bohrer, G. Hydrodynamic trait coordination and cost–benefit trade-offs throughout the isohydric–anisohydric continuum in trees. *Ecohydrology* **2019**, *12*, e2041. [CrossRef]

39. Tilman, D.; El Haddi, A. Drought and biodiversity in grasslands. *Oecologia* **1992**, *89*, 257–264. [CrossRef] [PubMed]

40. Symstad, A.J.; Chapin, F.S.; Wall, D.H.; Gross, K.L.; Huenneke, L.F.; Mittelbach, G.G.; Peters, D.P.; Tilman, D. Long-term and large-scale perspectives on the relationship between biodiversity and ecosystem functioning. *Bioscience* **2003**, *53*, 89–98. [CrossRef]

41. Petrie, M.D.; Bradford, J.B.; Hubbard, R.M.; Lauenroth, W.K.; Andrews, C.M.; Schlaepfer, D.R. Climate change may restrict dryland forest regeneration in the 21st century. *Ecology* **2017**, *98*, 1548–1559. [CrossRef]

42. Walker, B.H.; Emslie, R.H.; Owen-Smith, R.N.; Scholes, R.J. To cull or not to cull: Lessons from a southern African drought. *J. Appl. Ecol.* **1987**, *24*, 381–401. [CrossRef]

43. Hitchcock, R.K. Coping with Uncertainty: Adaptive Responses to Drought. In *Sustainable Livelihoods in Kalahari Environments: A Contribution to Global Debates*; Oxford Geographical and Environmental Studies; Oxford University Press: Oxford, UK, 2002; pp. 161–171.

44. Conradie, S.R.; Woodborne, S.M.; Cunningham, S.J.; McKechnie, A.E. Chronic, sublethal effects of high temperatures will cause severe declines in southern African arid-zone birds during the 21st century. *Proc. Natl. Acad. Sci. USA* **2019**, *116*, 14065–14070. [CrossRef]

45. Suzman, J. *Affluence without Abundance: The Disappearing World of the Bushmen*; Bloomsbury Publishing: New York, NY, USA, 2017.

46. Hermans-Neumann, K.; Priess, J.; Herold, M. Human migration, climate variability, and land degradation: Hotspots of socio-ecological pressure in Ethiopia. *Reg. Environ. Chang.* **2017**, *17*, 1479–1492. [CrossRef]

47. Hummel, D. Climate change, land degradation and migration in Mali and Senegal—some policy implications. *Migr. Dev.* **2015**, *5*, 211–233. [CrossRef]

48. Buhaug, H.; Nordkvelle, J.; Bernauer, T. One effect to rule them all? A comment on climate and conflict. *Clim. Chang.* **2014**, *127*, 391–397. [CrossRef]

49. Xu, J. Sand-dust storms in and around the Ordos Plateau of China as influenced by land use change and desertification. *Catena* **2006**, *65*, 279–284. [CrossRef]

50. Al-Awadhi, J.M.; Al-Dousari, A.M.; Khalaf, F.I. Influence of land degradation on the local rate of dust fallout in Kuwait. *Atmos. Clim. Sci.* **2014**, *4*, 437–446. [CrossRef]

51. De la Paix, M.J.; Lanhai, L.; Xi, C.; Ahmed, S.; Varenyam, A. Soil degradation and altered flood risk as a consequence of deforestation. *Land Degrad. Dev.* **2013**, *24*, 478–485. [CrossRef]

52. Paeth, H.; Born, K.; Girmes, R.; Podzun, R.; Jacob, D. Regional climate change in tropical and northern Africa due to greenhouse forcing and land use changes. *J. Clim.* **2009**, *22*, 114–132. [CrossRef]

53. Arribas, A.; Gallardo, C.; Gaertner, M.; Castro, M. Sensitivity of the Iberian Peninsula climate to a land degradation. *Clim. Dyn.* **2003**, *20*, 477–489. [CrossRef]

54. Bonn Challenge. Available online: www.bonnchallenge.org (accessed on 12 February 2020).

55. Sankaran, M.; Hanan, N.P.; Scholes, R.J.; Ratnam, J.; Augustine, D.J.; Cade, B.S.; Gignoux, J.; Higgins, S.I.; Le Roux, X.; Ludwig, F.; et al. Determinants of woody cover in African savannas. *Nature* **2005**, *438*, 846–849. [CrossRef]

56. Bond, W.J.; Stevens, N.; Midgley, G.F.; Lehmann, C.E. The trouble with trees: Afforestation plans for Africa. *Trends Ecol. Evol.* **2019**, *34*, 963–965. [CrossRef]

57. Allison, G.B.; Gee, G.W.; Tyler, S.W. Vadose-zone techniques for estimating groundwater recharge in arid and semiarid regions. *Soil Sci. Soc. Am. J.* **1994**, *58*, 6–14. [CrossRef]

58. Sternberg, T. Piospheres and pastoralists: Vegetation and degradation in steppe grasslands. *Hum. Ecol.* **2012**, *40*, 811–820. [CrossRef]

59. Briske, D.D.; Derner, J.D.; Brown, J.R.; Fuhlendorf, S.D.; Teague, W.R.; Havstad, K.M.; Gillen, R.L.; Ash, A.J.; Willms, W.D. Rotational grazing on rangelands: Reconciliation of perception and experimental evidence. *Rangel. Ecol. Manag.* **2008**, *61*, 3–17. [CrossRef]

60. Farmer, H. Understanding Impacts of Water Supplementation in a Heterogeneous Landscape. Ph.D. Thesis, University of the Witwatersrand, Johannesburg, South Africa, 2010.

61. Cornelis, W.; Waweru, G.; Araya, T. Building Resilience against Drought and Floods: The Soil-Water Management Perspective. In *Sustainable Agriculture Reviews*; Springer: Cham, Switzerland, 2019; Volume 29, pp. 125–142.

62. O'Farrell, P.J.; Anderson, P.M.L.; Milton, S.J.; Dean, W.R.J. Human response and adaptation to drought in the arid zone: Lessons from southern Africa. *S. Afr. J. Sci.* **2009**, *105*, 34–39.

63. Hobbs, R.J. Landscapes, ecology and wildlife management in highly modified environments—An Australian perspective. *Wildl. Res.* **2005**, *32*, 389–398. [CrossRef]
64. Wallander, S.; Aillery, M.; Hellerstein, D.; Hand, M. The role of conservation programs in drought risk adaptation. *Econ. Res. Serv. Econ. Res. Rep.* **2013**, *148*, 1–68.
65. Young, R. Beyond the Year of the Outback: What now for rural Australia? *Impact* **2003**, 8.
66. Carruthers, J. Wilding the farm or farming the wild? The evolution of scientific game ranching in South Africa from the 1960s to the present. *Trans. R. Soc. S. Afr.* **2008**, *63*, 160–181.

Review

Integrating Agriculture and Ecosystems to Find Suitable Adaptations to Climate Change

Anastasia W. Thayer [1,*,†], Aurora Vargas [1,†], Adrian A. Castellanos [2,†], Charles W. Lafon [3], Bruce A. McCarl [1], Daniel L. Roelke [4], Kirk O. Winemiller [2] and Thomas E. Lacher [2,5]

[1]   Department of Agricultural Economics, Texas A&M University, College Station, TX 77843, USA; avarga5@tamu.edu (A.V.); mccarl@tamu.edu (B.A.M.)

[2]   Department of Wildlife and Fisheries Sciences, Texas A&M University, College Station, TX 77843, USA; acastellanos@tamu.edu (A.A.C.); k-winemiller@tamu.edu (K.O.W.); tlacher@tamu.edu (T.E.L.)

[3]   Department of Geography, Texas A&M University, College Station, TX 77843, USA; clafon@geog.tamu.edu

[4]   Department of Marine Biology, Texas A&M University, Galveston, TX 77554, USA; droelke@tamu.edu

[5]   Center for Coffee Research and Education, Texas A&M University, College Station, TX 77843, USA

*   Correspondence: anastasia.thayer@usu.edu; Tel.: +1-4357970793

†   Seniority of authorship shared by first three authors.

Received: 9 December 2019; Accepted: 8 January 2020; Published: 9 January 2020

Abstract: Climate change is altering agricultural production and ecosystems around the world. Future projections indicate that additional change is expected in the coming decades, forcing individuals and communities to respond and adapt. Current research efforts typically examine climate change effects and possible adaptations but fail to integrate agriculture and ecosystems. This failure to jointly consider these systems and associated externalities may underestimate climate change impacts or cause adaptation implementation surprises, such as causing adaptation status of some groups or ecosystems to be worsened. This work describes and motivates reasons why ecosystems and agriculture adaptation require an integrated analytical approach. Synthesis of current literature and examples from Texas are used to explain concepts and current challenges. Texas is chosen because of its high agricultural output that is produced in close interrelationship with the surrounding semi-arid ecosystem. We conclude that future effect and adaptation analyses would be wise to jointly consider ecosystems and agriculture. Existing paradigms and useful methodology can be transplanted from the sustainable agriculture and ecosystem service literature to explore alternatives for climate adaptation and incentivization of private agriculturalists and consumers. Researchers are encouraged to adopt integrated modeling as a means to avoid implementation challenges and surprises when formulating and implementing adaptation.

Keywords: climate change; adaptation; agriculture; ecosystems; externalities

1. Introduction

Since the first World Climate Conference in 1979, researchers have been able to document and quantify the effects of anthropogenic climate change on physical climate, human, and natural systems [1]. Due to the economic importance of agriculture and strong ties to ecoregion diversity [2], plus the emergence of dramatic, recent climate change [3], Texas is an ideal region in which to assess the effects of climate change across natural and managed systems. With large parts of Texas being classified as semi-arid, the warmer and more arid regional shifts caused by climate change will be especially critical in long term decision making and in developing adaptation strategies. Over the last 20 years, temperatures across Texas have increased by 0.5–1.5 °C [4]. Precipitation patterns statewide have shifted, with statewide precipitation in the last century exhibiting increased rainfall in the east and decreased rainfall in the west [5]. At the same time, extreme weather events such as heavy rains, extreme droughts, tropical storms, and hurricanes, are becoming more common [4]. As a result, individuals and landscapes are responding. Winter wheat is no longer flowering at the same time as it did historically [6]. There have been marked declines in the quality and amount of habitat for birds [6,7], mussels [8], and butterflies [9], among many other species. Changes in temperature, precipitation, atmospheric carbon dioxide concentrations, and extreme event frequency/severity are impacting the distribution and function of agriculture and ecosystems in Texas.

As global carbon and other greenhouse gas emissions continue to increase, future model projections suggest that additional climate alterations are inevitable. Furthermore, even if global emissions were to return to pre-industrial levels the atmosphere would stay at current concentrations for many decades, with global surface temperatures continuing to increase and physical climate effects persisting well beyond 2100 [10]. Under a scenario where global emissions continue to increase through 2100 (under Representative Concentration Pathway 8.5) as they have for the last 100 years or more, it is expected that temperatures will increase across Texas and the Great Plains of the United States with 2070–2099 minimum temperatures approximately 6–8 °C higher and maximum temperatures 5–7 °C higher relative to 1971–1999 [11]. Precipitation patterns are also expected to become more extreme with the number of consecutive dry days increasing in 2070–2099 by 0–18% and maximum one-day precipitation totals increasing by 12–18% again compared to 1971–1999 [11]. Furthermore, in Texas, the number of days over 38 °C is expected to increase as is the number of warm nights [12]. Future conditions are expected to lead to even more agricultural and ecosystems shifts, disruptions, and production variability.

Given the already observable current impacts of climate change and projections of inevitable larger impacts, understanding how systems will respond and adapt is critical to maintain function. Systems that rely on climate characteristics and atmospheric carbon dioxide are especially vulnerable. The rapid rate of change poses unprecedented threats [13]. This is especially true for agriculture and ecosystems as they are fundamentally reliant on climate and carbon dioxide for productivity and their mix of available products/services. Interdependencies in resource usage, competition for space, and the movement of water, nutrients, and species among agriculture and ecosystems lead in effect to a unified interdependent system facing common drivers and constraints.

To date, research efforts on climate change effects and possible adaptations have been largely independent, concentrating on either agriculture or non-agricultural ecosystems but not both simultaneously. Failure to jointly address the effects and inform on the consequences of adaptations generates only a partial view of vulnerabilities and the implications of possible adaptations. This work argues that evaluations at the intersection of agriculture and ecosystems allow for analysis of synergies, feedbacks, and tradeoffs. Analyses that integrate impacts on and responses by both human and natural systems create a more robust, complex, and holistic evaluation of climate-change-related threats and possible adaptive decision making.

Additionally, global analyses overlook important regional characteristics and peculiarities that color vulnerabilities and adaptation implications. Here, we draw together evidence on how climate change and possible adaptations affect agriculture and ecosystems both individually and in interaction. We ask what the research and policy implications of analyses that ignore linkages between agriculture and ecosystems when exploring climate change effects and adaptation are. We find that a joint systems analysis is more informative as an input to ecosystem and agricultural management. In particular, in the Texas semi-arid setting we (1) briefly review the literature on the main impacts and vulnerabilities imposed on agriculture and ecosystems, (2) describe the interdependency of agriculture and ecosystems and the need for integrated climate change research, (3) discuss current and future adaptation possibilities and appraisal approaches, (4) introduce challenges for research in general and in the Texas-specific setting, and (5) argue the need for integrated research and modeling when understanding impacts of a climate-evolved future and the possibility of adaptation action.

2. Literature Review

Most climate change impacts research has considered agriculture and ecosystems to be independent of one another [14–17]. Few studies attempt to analyze the joint impact of adaptations or propose potential adaptation strategies that would reduce the negative impact of climate change across both systems. Recognizing the state of knowledge in each system, methodology used to date, and remaining research gaps would help identify mutually beneficial research needs and synergies. It would also contribute to an understanding of how addressing the systems jointly can help identify tradeoffs and possibilities for mutually beneficial outcomes. Here, we review the existing literature related to climate change impacts for agricultural crops, livestock, and food production along with general impacts on ecosystems, vegetation, and aquatic systems. A summary of climate change impacts on agriculture and ecosystems is shown in Table 1 and below we present a discussion for each system.

2.1. Agricultural Studies

Climate change is expected to have differing effects on cropping systems globally due to regionally specific physical conditions as well as differing mixes of crop and livestock types. For some crops in some regions, climate change has reduced current yields and is expected to reduce long-term agricultural productivity [18]. Agricultural research into climate change effects on crops has benefitted from observationally rich and geographically detailed datasets [19]. Studies can also take advantage of long-term, highly controlled, multi-site, manipulatable experimental studies [20,21]. The current extent of climate change has been shown to shift crop geographic distributions toward higher latitudes and elevations [19,22]. Studies have demonstrated that future crop productivity is expected to be limited by increased variability in weather and physical growing conditions [23], differentially impacted by carbon dioxide concentrations [24,25], and limited on a regional basis by dwindling water availability [26]. Other effects such as slowing technological progress [25,27] and increased pest damages and pesticide costs [28] are all expected to further alter productivity and costs. In Texas, a climate that is becoming warmer and drier with a greater probability of extreme events is expected to lead to declining yields for crops such as cotton [29] with greater variability due to extreme weather events [30]. Furthermore, lower soil moisture [31] is expected to increase aquifer pumping, in turn increasing drawdown and water stress. Finally, increased pest, disease, and invasive species frequency is expected to raise management costs [28,32].

Livestock, especially cattle, are expected to be directly impacted by climate change and increased heat stress but also indirectly though impacts on forage and feed grain yield reductions [33]. In the US, direct livestock losses due to heat stress are estimated to be $2.4 billion annually from decreased reproduction rates, feed consumption, and feed efficiency affecting animal growth rates [34–37]. Lower forage and feed quality are also expected [38] as increased temperatures negatively affect growth conditions and nutrient availability [39]. In Texas, warmer and drier conditions are expected to reduce total livestock production through lower stocking rates and reduced per animal production. Lower grassland growth rates and nutritional quality will force increased supplemental feeding and costs [40]. Total grassland productivity is also expected to decrease with the expansion of woody plants [40], although movements of land from cropping are expected to increase grass land quantity [41]. Expansion and greater incidence of disease, ectoparasites, and other pests are expected to decrease animal productivity [30,40,41].

The impact of climate change on agricultural systems also has implications for land prices, transportation, storage, food safety, labor, and consumer prices. These critical processes within the supply chain for agricultural products are expected to be affected with alterations occurring at every stage of production, including input sourcing, packaging, and processing [42]. It has been suggested that additional precautions might have to be considered to maintain food safety and reduce spoilage, such as increased storage and cooling facilities [43,44]. Relevant to Texas, shifting US production capacity is expected to change routes and methods used to transport agricultural products [45]. On a larger scale, global shifts in production capacity are anticipated which will alter comparative advantages, international trading routes, and partnerships [42]. Agricultural prices will also be impacted. However, determining the direction, magnitude, and associated changes to producer and consumer welfare as a result is complicated [46–48]. For example, price changes may impact urban versus rural consumers or other sub-groups within the same market differently [48]. Agricultural labor supply is also predicted to be impacted, and, with it, rural incomes [49,50]. Finally, changes in agricultural land values are anticipated as historic land use is expected to shift either due to changes in the agricultural activity utilizing the land, land moving out of agricultural production all together, and/or changing values of land based on water or other resource availability [48].

2.2. Ecological Studies

While agricultural scientists have been able to study effects using large public datasets and publicly funded experiments, work on ecosystem effects has proven more difficult. For natural ecosystems there is a lack of widespread data availability. Furthermore, most available data sets focus on one species and/or geographic area with substantial inconsistencies in study–time horizon [51,52]. Nevertheless, consistent impacts and vulnerabilities have been identified.

Foremost, biodiversity is threatened by climate change due to the rising trend and magnitude of change over a short timeline. This impacts all levels of biodiversity, from individual organisms to populations and ecosystems [53]. Extirpation of regional populations and global extinction continue to be the most visible impacts, although establishing the extent of climate change causality remains challenging because species vary in their capacity to adapt [54]. However, in recent years, our ability to model these shifts has improved due to the creation and continued proliferation of biodiversity data repositories (e.g., GBIF) and VertNet, etc.; [55]) and finer scale environmental data (e.g., EarthEnv, SoilGrids, and WorldClim; [56–58]), in addition to improvements in climate model resolution [59].

Climate change has already been found to alter species geographic distribution, phenology, behavior, and patterns of habitat use, with more change expected in the coming decades. Organisms adapt to inhospitable physical climate conditions by shifting, expanding, or contracting their historic ranges [60,61], and for a few species, perishing. Climate-associated range shifts have been observed across a wide geographic and taxonomic scope, including flora in the Himalayas [62] and the western United States [63], birds in New Guinea [64], Amazonian fish [65], and small mammal communities [66], just to name a few. As an example, Parmesan and Yohe [51] sampled 1598 species across multiple taxa, of which 59% had exhibited changes in their phenology or distribution over the past 20 to 140 years. Furthermore, the presence of novel climate niches and geographic barriers that reduce dispersal and gene flow [67] will likely limit the potential for natural adaptation.

Across Texas, species already are showing dramatic responses to climate change. For example, migration patterns for resident birds have been impacted [68]. Model projections indicate that some rodent species will go extinct and species geographic ranges are expected to shift 54% or more depending on the extent of climate change [69]. Diseases, invasive species, and pests are expected to change their distribution with ecological consequences, with, for example, tick vectors shifting and likely bringing diseases into new regions, impacting both humans and wildlife [70,71]. This will require more complex eradication and control strategies [72] as tick-borne disease relationships are changed [73]. These are just a few examples of the profound impact on wildlife populations.

Vegetation communities are also responding. Plants are governed by stress and disturbance, and climate-induced changes to these factors will alter vegetation composition, productivity, and distribution [74,75]. Changes in temperature, precipitation, and climatic extremes can increase stress and limit plant growth [76,77]. In Texas and other semi-arid regions, warming-induced increases in evapotranspiration are expected to reduce plant productivity [78,79]. Moreover, as precipitation variability increases, grassland productivity decreases regardless of constant average rainfall [80]. Clearly, climate variability matters when considering damage from climate change. Shifts in disturbance regimes are expected due to changes in the prevalence and distribution of fires, floods, hurricanes, and insect outbreaks, thus forcing communities into altered states [81,82]. These transitions can occur rapidly when severe disturbances are combined with increasing stress and they can lead to permanent vegetation community changes [83,84].

Table 1. A summary of climate change impacts on agriculture and ecosystems.

	Climate Impacts on Agriculture	Citations
Crops	Crop mixes and distributions are shifting northward to higher elevation	[19,22]
	Future crop productivity (1) limited by increased variability in weather and physical growing conditions, (2) differentially impacted by carbon dioxide concentrations, (3) limited by dwindling water availability, (4) limited by slowing technological progress, and (5) limited by increased pesticide costs.	[23–28]
	Texas: warmer and drier climate-reduced crop yields and increased losses due to extreme weather events	[29,30]
	Texas: lower soil moisture leading to increased aquifer pumping and water stress	[31]
	Texas: increased frequency of pest, disease, and invasive species which raises crop management costs	[28,32]
Livestock	Increased heat stress and reduced forage and feed growth	[33]
	Livestock losses from decreased reproduction rates, feed consumption, and feed efficiency affecting animal growth rates	[34–37]
	Lower forage and feed quality due to increased temperatures affecting growth and nutrient availability	[38,39]
	Texas: lower stocking rates and reduced per animal production due to warmer and drier conditions	[33]
	Texas: increased supplemental feeding due to lower grassland growth rates, quality, and acreage with the expansion of woody plants	[40]
	Texas: decreased animal productivity due to the expansion and greater incidence of disease, ectoparasites, and other pests	[30,40,41]
Supply Chain	Input sourcing, packaging, and processing affected by climate change	[42]
	Additional storage and cooling facilities necessary to maintain food safety and reduce spoilage from increased temperatures	[42–44]
	Shifting US production capacity will change transportation routes and methods	[45]
	Altered comparative advantages, international trading routes, partnerships, and trade agreements due to shifts in production	[42]
	Difficulty in determining the direction, magnitude, and associated changes to producer and consumer welfare	[46–48]
	Agricultural labor supply is predicted to be impacted, and with it rural incomes	[49,50]
	Changes in agricultural land values as historic land use shifts	[48]
	Climate Impacts on Ecosystems	**Citations**
Fauna	Biodiversity is threatened due to the trend and magnitude of rapid changes over a short timeline	[53]
	Extirpation due to varied capacity of species to adapt to environmental changes brought about by climate change	[54]
	Organisms respond to inhospitable physical climate conditions by shifting, expanding, or contracting their historic ranges	[60,61]
	Barriers to dispersal that reduce gene flow in landscapes which limit potential for natural adaptation	[67]
	Texas: migration patterns for resident birds have been impacted	[68]
	Texas: some rodent species will go extinct and geographic shifts of 54% or more will occur	[69]
	Texas: tick vectors are shifting and will likely bring diseases into new regions impacting humans and wildlife, resulting in more complex eradication and control strategies	[70–73]

Table 1. *Cont.*

	Climate Impacts on Agriculture	Citations
Flora	Altered vegetation composition, productivity, and distribution due to climate-induced stress and disturbance	[74,75]
	Limited plant growth due to changes in temperature, precipitation, or the incidence of climatic extremes	[76,77]
	Texas: reduced plant productivity due to increasing evapotranspiration	[78,79]
	Altered prevalence and distribution of fires, floods, hurricanes, and insect outbreaks forces communities into a stressed state which can lead to permanent changes to vegetation	[81–84]
Aquatic and Riparian	Hydrological environment areas that cycle nutrients, maintain water quality, and moderate lifecycle events such as spawning and recruitment are disrupted by climate changes	[85–88]
	Dewatered channel segments leading to habitat fragmentation due to reduced flows	[86,89]
	Texas: disrupted productivity and biodiversity of stressed freshwater inflows due to human appropriation	[90]
	Increased algal blooms due to warmer water temperatures and changes in rainfall	[91,92]

Aquatic and riparian systems are also affected. The hydrologic environment adds an additional layer of complexity as it also cycles nutrients, alters water quality [85–87], and moderates lifecycle events such as spawning and recruitment [88]. When rivers and streams in arid and semi-arid regions experience severely reduced flows, channel segments may become dewatered, resulting in habitat fragmentation and threatening the population viability of rare endemic species at scales that often extend well beyond the impacted habitat [86,89]. In Texas, freshwater inflows that support coastal ecosystems are expected to come under increasing stress from human appropriation and altered flow levels, and this will further disrupt future productivity and erode native biodiversity [90]. Looking ahead, harmful algal blooms are expected to become an increasing problem [91,92]. As conditions become drier and the magnitude and frequency of freshwater inflows decline, such algal blooms are likely to cause larger fish kills and substantial financial damages.

2.3. Summary Table

In Table 1, below, is a summary of the relevant literature from agriculture and ecosystem studies.

3. Need for an Integrated Approach

The above material clearly shows that climate change disruptions to temperature, precipitation, and extreme events threaten the health, function, and productivity of agriculture and more generally ecosystems. However, gaps remain in understanding and projecting future impacts, especially since critical interactions between agriculture and ecosystems have largely been ignored. These interactions can include externalities or unintended effects, additional drivers, feedback loops, and tradeoffs. For example, pesticide use is expected to increase as a result of emerging and expanding pest populations [28]. Pesticides impact not only ecosystems where they are applied but also have far reaching effects when they are transported via runoff and infiltration [93,94]. As another example, the interactions between cattle and grassland production and forage quality has not been well integrated into climate change research [41,95]. Warmer and drier climates stress livestock [34] but estimates of damages have not fully considered the additional effects of decreased shade cover and less water availability on rangeland grazing animals. Adaptation to such simultaneous stressors may lead to increased costs, lower productivity, and less revenue [30]. In Texas and other areas with extensive rangeland acreage and cropland under input-intensive agriculture, if synergistic impacts across the ecosystem are not considered, the costs of both projected and realized climate change might be severely underestimated, leading to reduced adaptation action.

An improved understanding of how the systems interact and of the relevant feedbacks need to be developed. Arguably, the most widespread effort to begin to unify agriculture and ecosystems has occurred through monetary valuations of ecosystem services [96–99]. However, this effort falls short of holistically incorporating ecosystems and agricultural regimes into a shared conceptualization of climate change effects. Rather, ecosystem valuation is typically reduced to a short-sighted service value or a dynamic financially-discounted contribution over time [97,99]. When efforts to evaluate agricultural practices in the presence of ecosystems do take place, the results confirm that ignoring this duality leads to severely biased findings [100,101].

For climate change research, this forces a discussion of the validity of findings on climate change effects and adaptations when one system is analyzed in isolation. Some researchers have identified the need for integrated research and have presented loose guidelines or examples for how to merge studies, disciplines, and research priorities [100,102]. However, this is challenging, as the inclusion of increasing degrees of climate change exacerbates existing data limitations, timescale mismatches, geographic scale, unstudied but associated phenomena, and a need for adaptation action to avoid severe consequences. At its base, what is missing from much of current climate change research is an understanding of how rapid change affects the linkages between agriculture and ecosystems and in turn how resilience, future output, and, ultimately, the survival of communities, will be impacted.

Improved understanding could begin with an analysis of ecosystem services including agricultural and other markets as a service and conceptualization of regional interactions. The analysis could start by building off the framework given in Figure 1. Identification and understanding of ecosystem services provided to a particular agricultural system, market, or regional society could rely on existing ecosystem service literature [96–99,102,103]. Given that tradeoffs are known to exist between market services (crops, livestock, and water) and non-market ecosystem services (regulating, supporting, and cultural), attempts are needed to minimize negative impacts or assess best practices. Such examinations will enhance understanding of the problem, thereby avoiding incomplete analysis and flawed results. Identifying tradeoffs of overconsumption will be particularly important. Research efforts that strive to incorporate both systems into an analytic structure will provide more robust and broadly applicable findings and recommendations.

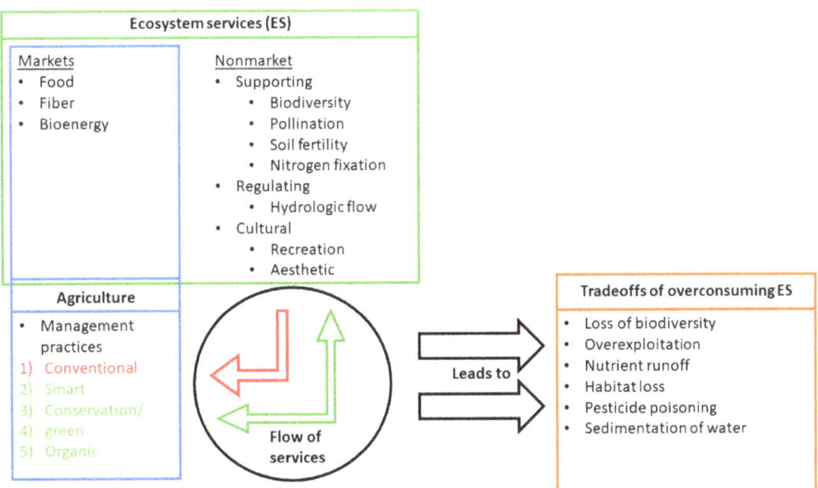

Figure 1. Joint ecosystem and agriculture modeling framework highlighting integrated nature of systems [96,103].

4. Sustainable Adaptation Challenges and Solutions

Agriculture and ecosystems are already reacting and responding to climate change by exhibiting altered productivity and species populations. In managed systems individuals, farmers, and researchers are trying to anticipate future changes and adapt in beneficial and cost-effective manners. In unmanaged systems, natural adaptation is occurring but not always in desirable ways, and in these cases management intervention is being contemplated often with incomplete knowledge of the consequences. Research efforts projecting effects and evaluating adaptation actions have an even greater incentive to consider an integrated agriculture and ecosystems framework.

Most often, agriculture and ecosystems adapt to climate change effects in response to altered physical climate. A flow chart showing how changes to physical climate can motivate adaptation can be seen in Figure 2. From previous examples, it is clear that changes to physical climate such as increased temperature and altered rainfall can change cereal crop yields [22]. In response, farmers adapt to altered yields by changing crop mixes. Similarly altered species abundance occurs when temperature rainfall or extremes affect regional ecosystems [53,54].

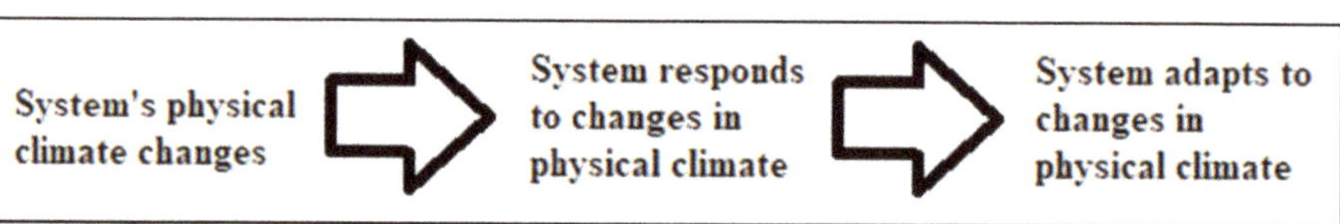

Figure 2. Flow chart describing the primary motivation for systems to adapt to climate change effects.

To date, agricultural systems have largely adapted or considered adapting to climate change through altering management practices and implementing policies that motivate desired behavior or management practices. Much of the published literature on agricultural adaptation uses a large geographic study area and shows that adaptations can lessen the negative impacts of future climate alterations. Within cropping systems, strategies proposed have included earlier planting times, changing crop mixes, and complete shifts of land out of crop into pasture [19,41,104]. For livestock, proposals include adopting more heat-tolerant breeds or species, changing stocking rates, providing shade or water, altering pest management, and shifting grazing seasons [40,105,106]. Overall, markets and other mechanisms for insurance, such as water rights regimes, water markets, and crop insurance, might also have to evolve, expand, or be redefined under climate change in order to mitigate risk to users [107,108].

As a general economic rule, undertaking adaptation strategies and supporting governmental or institutional efforts will only occur if they are judged superior to current practices [109]. The adaptation strategies listed above seek to minimize risk for individual producers but are usually considered to ignore potential externalities or impacts on the surrounding ecosystem (An externality is simply a market failure where the price of a good does not reflect its value.). In other words, while the chosen adaptations are efficient for a private producer, they may not be efficient for society, or cause substantial ecosystem damages. This commonly happens when looking at environmental goods and services because it is difficult to price all possible benefits and costs, which leads to a poor estimate of value. As seen in Equation (1), only when the marginal private cost (*MPC*) and marginal social cost (*MSC*) of an adaptation equals the marginal benefit (*MB*) do we see an efficient market and an accurately valued good [110], or in this case, a holistic adaptation strategy.

$$MB = MPC + MSC \tag{1}$$

When the *MSC* is not considered, the value of the externality is equal to the value of the *MSC*, thus making the adaptation inefficient. Inefficient adaptation and maladaptation could cause long-term damages that limit or slow beneficial adaptation to climate change, and, ultimately, increase the damages and costs of climate change [109,111]. Studies to date show that some adaptation strategies for agriculture could cause both winners and losers [112]. Thus, failure to incorporate the positive and negative externalities associated with adaptation efforts to a modeling framework in either agriculture or ecosystems has the potential to bias the estimates of benefits and costs causing poor or at least sub-optimal choices to be made.

Table 2 seeks to provide agricultural and ecosystem examples of potential externalities resulting from adaptations. Column 1 displays the underlying climate change stressor effect that is stimulating adaptation. Columns 2 and 3 show examples of corresponding agricultural management or ecosystem responses. It is useful to note that multiple adaptations can result from one climate change driver. For example, increased drought frequency and duration can motivate farmers to adapt by changing crop mix while also causing ecosystems to adapt, resulting in shifting vegetation mix and water retention.

Table 2. Eight examples of how climate stressors lead to adaptation in agriculture and the corresponding response in ecosystems which can be a positive or negative externality. The opposite case of adaptations in ecosystems and responses in agriculture is then presented.

Climate Stressor	Agricultural Adaptation	Ecosystem Service Externality
1. Increased temperature and drought	Diversifying livestock species [113–115]	Altered plant biodiversity and productivity [116–118]
	Crop land shift to grazing [19,41,119]	Increased root production in upper soil levels and carbon sequestration [120,121]
2. Increased temperature	Heat-tolerant animal breeds [103]	Dilution of disease prevalence [103,122,123]
3. Increased drought	Changing crop mix and rotation [19,124,125]	Increased soil quality [126,127]
Climate Stressor	**Ecosystem Adaptation**	**Agriculture System Externality**
4. Shifts in temperature and rainfall patterns	Land vegetative change and habitat fragmentation [128,129]	Reduction in pollinators and pollination [100,130–132]
5. Increased temperature	Reduced animal body size [133]	Altered diets and rangeland economic productivity such as stocking rates [41,114,134]
6. Increased drought	Shift in vegetation productivity and water retention [135,136]	Altered water supply and increased demand for irrigation [137,138]
7. Increased temperature and altered rainfall	Shifting species distribution [139,140]	Increased pesticide and herbicide costs [28,141,142]
8. Increased water temperature	Change in phenology [143,144]	Reduced fish survival [145–149]

These examples show that adaptation efforts or actions experienced in one system have spillover effects into the other which may impact the function and economic viability of the opposing system. These spillovers increase the uncertainty of outcomes. For researchers and policymakers seeking to make informed adaptation decisions and recommendations, simultaneous human and natural adaptation makes analysis and modeling efforts complex. At the same time, not all observed externalities are negative or are expected to increase damages from climate change. An agricultural adaptation such as switching to more heat-tolerant livestock breeds can introduce less competent hosts for pathogen transmission, diluting disease prevalence in ecosystems. Therefore, to accurately predict, respond to, and make recommendations for adaptation strategies in response to future climate change, a framework that includes both ecosystems and agriculture must be adopted.

Solutions: Improving Modeling Efforts

Based on the literature, three avenues of approach arise that can assist modeling efforts in merging ecosystems and agriculture systems. These are (1) incorporating alternative practices that can lessen the impact of agriculture on ecosystems, (2) incorporating and advancing modeling of ecosystem services and the way they are affected by agricultural activity, and (3) modeling means of providing economic incentives to encourage adoption of conservation or environmental policies.

Practices exist that can be adopted to reduce agricultural impacts on surrounding ecosystems, as shown by the long US history with soil conservation [150]. There is a large body of literature that champions conservation agriculture practices such as reduced/no tillage, retention of crop residues, and altered crop rotation that benefit ecosystem services regulation and provisioning, via expanded carbon sequestration, reduced erosion, and improved soil health [151,152]. Other studies support expanding organic agriculture, which has been shown to reduce off-farm impacts while preserving ecosystem services [103]. However, while these strategies reduce agricultural impacts to ecosystems, more research is needed to optimize management practices so that organic yields can consistently meet or exceed the yields of conventional agriculture since population growth is increasing food demand [153]. Additionally, precision agriculture and climate smart agriculture might also offer a solution, as they rely on optimizing current conventional agricultural techniques and responding to climate change at a more localized level [154–157]. The benefit of these approaches is that they inherently have characteristics which benefit the environment. Continued research and efforts toward incorporating positive environmental externalities into production agriculture decision-making above could present alternative ways to reframe the current narrative to benefit health of both agriculture and ecosystems.

Ecosystem service analysis and modeling methodologies present potential solutions for integrating ecosystems and agriculture. Non-market valuation of ecosystem services determines a value for a particular facet of the ecosystem which benefits humans and provides a quantitative measurement for use in adaptation strategy and associated policy analysis [98]. This approach can be useful for valuing ecosystems and placing them on an equal footing with market transactions. Tools such as the Integrated Valuation of Ecosystem Services and Tradeoffs (InVEST) models help give a quantitative value to ecosystem services in an easy, open-source platform which can then be used to look at land-use change or other scenarios [97]. Many of these modeling efforts will rely on ambitious data collecting efforts such as National Ecological Observatory Network (NEON), a 30-year long-term research project designed to capture a wide range of ecosystem process indicators [158]. These data can then be integrated with output from other systems to incorporate ecosystems, their processes, and how they change over time into decision analysis [159]. More recently, many of the machine learning techniques typically used in species distribution modeling [160] are being used to model biodiversity [82], agricultural suitability [161], and crop mix shifts [162]. As the use of these methods spreads, they will be able to help characterize the contributions and sensitivities of ecosystems relevant to agriculture. Creating causal networks and truly assimilating ecosystems and agriculture could benefit from ecosystem service valuation research and additional long-term ecosystem monitoring databases.

Finally, policies that incentivize private landowner environmental efforts could aid in the public's realized benefits from healthier ecosystems. Also, monitoring the results of such programs would increase understanding of how ecosystems and agriculture interact. In Texas, specifically, most of the land is privately owned and programs such as Texas Ecological Laboratories ("Texas Ecolab") can facilitate connections between researchers and landowners. Such an exercise generates data while advancing environmental goals such as conservation and the conduct of research that improves state interests [163,164]. It also economically rewards cooperating landowners for their efforts [165]. Fostering partnerships with private landowners is especially beneficial, as oftentimes there is distrust of government entities [166] that hampers conservation or research efforts [167].

5. Conclusions

Future impacts of climate change are inevitable, stimulating ecosystem and agriculture impacts and responses. Understanding how ecosystems and agriculture are inherently linked and projecting climate change consequences, and then using this knowledge to inform adaptation actions has the potential to improve policy and lower social/environmental impacts. Moreover, such an understanding will further protect against inefficient and even detrimental adaptations that cause long-term disruptions. Developing an integrated framework that jointly considers both agriculture and ecosystems enhances our knowledge of the inherent tapestry of occurring interactions. Understanding these interactions will help maintain and enhance system resiliency necessary to produce food and human livelihoods while maintaining a productive, high-quality environment.

While an integrated ecosystem and agricultural framework is recommended, other issues remain that will challenge our ability to adapt to climate change in a way that minimizes damages for future generations as well as ecosystems. Firstly, large-scale studies and solutions must be paired with local and regional analysis, interpretation, and flexible implementation to avoid missing localized phenomena [30,40,168,169]. Action to integrate agriculture and ecosystems will reveal knowledge gaps related to externalities, feedbacks, and dynamics within and between systems. Secondly, while this work motivates the need for an integrated research framework, we could not find a specific example of where an integrated model proved superior to disjoint efforts. Thirdly, while initially difficult to overcome, increased monitoring and identification of critical data needs will contribute to resolving these challenges. Future research could address identifying and quantifying cause and effect relationships among systems and some research efforts could focus on case studies showing the added benefit of additional data and integrated analyses. Fourthly, funding is also somewhat compartmentalized to individual areas and the development of broader funding opportunities to support this research is needed.

Overall, while an integrated ecosystem and agricultural framework will not solve all climate change challenges, it might help remove some of the uncertainty [170], balance conflicting objectives, and present more nuanced solutions to a complex problem.

Author Contributions: Conceptualization was carried out by A.W.T., A.V., and A.A.C., as well as T.L. and B.A.M.; methodology, same as previous; formal analysis was completed by A.W.T., A.V., and A.A.C.; investigation was formulated by all authors; writing—original draft preparation was carried out by A.W.T. with contributions from A.V. and A.A.C.; review and editing were performed by all authors with substantial contributions by A.W.T., A.V., A.A.C, T.E.L., and B.A.M.; supervision, T.E.L. and B.A.M.; project administration, T.E.L.; funding acquisition, T.E.L. and B.A.M. All authors have read and agreed to the published version of the manuscript.

Funding: This research was funded by the Texas A&M Grand Challenges Proposal Competition "Building Climate Resilience: Seeking Sustainable Solutions for Water, Agriculture and Biodiversity in Arid Regions".

Acknowledgments: The authors would like to thank the contributions of Professor David Briske as well as participants and conference speakers at the events related to the Texas A&M Grand Challenges Proposal Competition "Building Climate Resilience: Seeking Sustainable Solutions for Water, Agriculture and Biodiversity in Arid Regions".

Conflicts of Interest: The authors declare no conflict of interest.

References

1. Zillman, J.W. A history of climate activities. *WMO Bull.* **2009**, *58*, 141.
2. Ricketts, T.H.; Dinerstein, E.; Olson, D.M.; Loucks, C.J.; Eichbaum, W.; DellaSala, D.; Kavanagh, K.; Hedao, P.; Hurley, P.T.; Carney, K.M.; et al. *Terrestrial Ecoregions of North America: A Conservation Assessment*; Island Press: Washington, DC, USA, 1999.
3. Schmidly, D.J. *Texas Natural History: A Century of Change*; Texas Tech University Press: Lubbock, TX, USA, 2002.
4. U.S. Environmental Protection Agency. What Climate Change Means for Texas. Available online: https://www.epa.gov/sites/production/files/2016-09/documents/climate-change-tx.pdf (accessed on 25 September 2019).
5. U.S. Environmental Protection Agency. Climate Change Indicators: Heavy Precipitation. Available online: https://www.epa.gov/climate-indicators/climate-change-indicators-heavy-precipitation (accessed on 25 September 2019).
6. Melillo, J.M.; Richmond, T.T.; Yohe, G. Climate change impacts in the United States. *Third Natl. Clim. Assess.* **2014**, *52*. [CrossRef]
7. Rittenhouse, C.D.; Pidgeon, A.M.; Albright, T.P.; Culbert, P.D.; Clayton, M.K.; Flather, C.H.; Masek, J.G.; Radeloff, V.C. Land-cover change and avian diversity in the conterminous United States. *Conserv. Biol.* **2012**, *26*, 821–829. [CrossRef]
8. Karatayev, A.Y.; Burlakova, L.E.; Miller, T.D.; Perrelli, M.F. Reconstructing historical range and population size of an endangered mollusc: Long-term decline of *Popenaias popeii* in the Rio Grande, Texas. *Hydrobiologia* **2018**, *810*, 333–349. [CrossRef]

9. Flockhart, D.T.T.; Pichancourt, J.B.; Norris, D.R.; Martin, T.G. Unravelling the annual cycle in a migratory animal: Breeding-season habitat loss drives population declines of monarch butterflies. *J. Anim. Ecol.* **2015**, *84*, 155–165. [CrossRef] [PubMed]

10. Diffenbaugh, N.S.; Stone, D.A.; Thorne, P.; Giorgi, F.; Hewitson, B.C.; Jones, R.G.; van Oldenborgh, G.J. 2014: Cross-chapter box on the regional climate summary figures. In *Climate Change 2014: Impacts, Adaptation, and Vulnerability. Part A: Global and Sectoral Aspects. Contribution of Working Group II to the Fifth Assessment Report of the Intergovernmental Panel on Climate Change*; Field, C.B., Barros, V.R., Dokken, D.J., Mach, K.J., Mastrandrea, M.D., Bilir, T.E., Chatterjee, M., Ebi, K.L., Estrada, Y.O., Genova, R.C., et al., Eds.; Cambridge University Press: Cambridge, UK; New York, NY, USA, 2014; pp. 137–141.

11. Wuebbles, D.J.; Kunkel, K.; Wehner, M.; Zobel, Z. Severe weather in the United States under a changing climate. *Eos Trans. AGU* **2014**, *95*, 149–150. [CrossRef]

12. Shafer, M.; Ojima, D.; Antle, J.M.; Kluck, D.; McPherson, R.A.; Petersen, S.; Scanlon, B.; Sherman, K. Great Plains. In *Climate Change Impacts in the United States: The Third National Climate Assessment*; Melillo, J.M., Richmond, T., Yohe, G.W., Eds.; USA Global Change Research Program: Washington, DC, USA, 2014; pp. 441–461. [CrossRef]

13. Williams, J.W.; Jackson, S.T.; Kutzbach, J.E. Projected distributions of novel and disappearing climates by 2100 AD. *Proc. Natl. Acad. Sci. USA* **2007**, *104*, 5738–5742. [CrossRef] [PubMed]

14. Kardol, P.; Reynolds, W.N.; Norby, R.J.; Classen, A.T. Climate change effects on soil microarthropod abundance and community structure. *Appl. Soil Ecol.* **2011**, *47*, 37–44. [CrossRef]

15. Chen, C.C.; McCarl, B.A. Hurricanes and Possible Intensity Increases: Effects on and Reactions From US Agriculture. *Am. J. Agric. Econ.* **2009**, *41*, 125–144. [CrossRef]

16. Schlenker, W.; Roberts, M.J. Nonlinear temperature effects indicate severe damages to U.S. crop yields under climate change. *Proc. Natl. Acad. Sci. USA* **2009**, *106*, 15594–15598. [CrossRef]

17. Lobell, D.B.; Hammer, G.L.; McLean, G.; Messina, C.; Roberts, M.J.; Schlenker, W. The critical role of extreme heat for maize production in the United States. *Nat. Clim. Chang.* **2013**, *3*, 497–501. [CrossRef]

18. Lobell, D.B.; Schlenker, W.; Costa-Roberts, J. Climate trends and global crop production since 1980. *Science* **2011**, *333*, 616–620. [CrossRef] [PubMed]

19. Cho, S.J.; McCarl, B.A. Climate change influences on crop mix shifts in the United States. *Sci. Rep.* **2017**, *7*, 40845. [CrossRef] [PubMed]

20. Camara, K.M.; Payne, W.A.; Rasmussen, P.E. Long-term effects of tillage, nitrogen, and rainfall on winter wheat yields in the Pacific Northwest. *Agron. J.* **2003**, *95*, 828–835. [CrossRef]

21. Leakey, A.D.; Uribelarrea, M.; Ainsworth, E.A.; Naidu, S.L.; Rogers, A.; Ort, D.R.; Long, S.P. Photosynthesis, productivity, and yield of maize are not affected by open-air elevation of CO_2 concentration in the absence of drought. *Plant Physiol.* **2006**, *140*, 779–790. [CrossRef] [PubMed]

22. Fei, C.J.; McCarl, B.A.; Thayer, A.W. Estimating the impacts of climate change and potential adaptation strategies on cereal grains in the United States. *Front. Ecol. Evol.* **2017**, *5*, 62. [CrossRef]

23. McCarl, B.A.; Villavicencio, X.; Wu, X. Climate change and future analysis: Is stationarity dying? *Am. J. Agric. Econ.* **2008**, *90*, 1241–1247. [CrossRef]

24. Long, S.P.; Ainsworth, E.A.; Leakey, A.D.; Nösberger, J.; Ort, D.R. Food for thought: Lower-than-expected crop yield stimulation with rising CO_2 concentrations. *Science* **2006**, *312*, 1918–1921. [CrossRef]

25. Attavanich, W.; McCarl, B.A. How is CO_2 affecting yields and technological progress? A statistical analysis. *Clim. Chang.* **2014**, *124*, 747–762. [CrossRef]

26. Chen, C.C.; Gillig, D.; McCarl, B.A. Effects of climatic change on a water dependent regional economy: A study of the Texas Edwards aquifer. *Clim. Chang.* **2001**, *49*, 397–409. [CrossRef]

27. Kapilakanchana, M. The Effect of Technological Progress, Demand, and Energy Policy on Agricultural and Bioenergy Markets. Ph.D. Thesis, Texas A&M University, College Station, TX, USA, 2016.

28. Chen, C.C.; McCarl, B.A. An investigation of the relationship between pesticide usage and climate change. *Clim. Chang.* **2001**, *50*, 475–487. [CrossRef]

29. Wang, Z. Three Essays on Climate Change, Renewable Energy, and Agriculture in the US. Ph.D. Thesis, Texas A&M University, College Station, TX, USA, 2018.

30. Steiner, J.L.; Briske, D.D.; Brown, D.P.; Rottler, C.M. Vulnerability of Southern Plains agriculture to climate change. *Clim. Chang.* **2018**, *146*, 201–218. [CrossRef]

31. Cook, B.I.; Ault, T.R.; Smerdon, J.E. Unprecedented 21st century drought risk in the American Southwest and Central Plains. *Sci. Adv.* **2015**, *1*, e1400082. [CrossRef] [PubMed]

32. Walthall, C.L.; Hatfield, J.; Backlund, P.; Lengnick, L.; Marshall, E.; Walsh, M.; Adkins, S.; Aillery, M.; Ainsworth, E.A.; Ammann, C.; et al. *Climate Change and Agriculture in the United States: Effects and Adaptation*; Technical Bulletin 1935; U.S. Department of Agriculture: Washington, DC, USA, 2012.

33. Rötter, R.; Van de Geijn, S.C. Climate change effects on plant growth, crop yield and livestock. *Clim. Chang.* **1999**, *43*, 651–681. [CrossRef]

34. St. Pierre, N.R.; Cobanov, B.; Schnitkey, G. Economic losses from heat stress by US livestock industries. *J. Dairy Sci.* **2003**, *86*, E52–E77. [CrossRef]

35. Ferreira, F.C.; Gennari, R.S.; Dahl, G.E.; De Vries, A. Economic feasibility of cooling dry cows across the United States. *J. Dairy Sci.* **2016**, *99*, 9931–9941. [CrossRef]

36. Yu, C.H.; Park, S.C.; McCarl, B.; Amosson, S.H. Feedlots, Air Quality and Dust Control-Benefit Estimation under Climate Change. In *2012 Annual Meeting, August 12–14, 2012, Seattle, Washington (No. 124736)*; Agricultural and Applied Economics Association: Milwaukee, WI, USA, 2012.

37. Yu, C.H. Case Studies on the Effects of Climate Change on Water, Livestock and Hurricanes. Ph.D. Thesis, Texas A&M University, College Station, TX, USA, 2014.

38. Derner, J.; Briske, D.; Reeves, M.; Brown-Brandl, T.; Meehan, M.; Blumenthal, D.; Travis, W.; Augustine, D.; Wilmer, H.; Scasta, D.; et al. Vulnerability of grazing and confined livestock in the Northern Great Plains to projected mid-and late-twenty-first century climate. *Clim. Chang.* **2018**, *146*, 19–32. [CrossRef]

39. Craine, J.M.; Elmore, A.; Angerer, J.P. Long-term declines in dietary nutritional quality for North American cattle. *Environ. Res. Lett.* **2017**, *12*, 044019. [CrossRef]

40. Briske, D.D.; Joyce, L.A.; Polley, H.W.; Brown, J.R.; Wolter, K.; Morgan, J.A.; McCarl, B.A.; Bailey, D.W. Climate-change adaptation on rangelands: Linking regional exposure with diverse adaptive capacity. *Front. Ecol. Environ.* **2015**, *13*, 249–256. [CrossRef]

41. Mu, J.E.; McCarl, B.A.; Wein, A.M. Adaptation to climate change: Changes in farmland use and stocking rate in the USA. *Mitig. Adapt. Strateg. Glob. Chang.* **2013**, *18*, 713–730. [CrossRef]

42. Porter, J.R.; Xie, L.; Challinor, A.J.; Cochrane, K.; Howden, S.M.; Iqbal, M.M.; Lobell, D.B.; Travasso, M.I. Food security and food production systems. In *Climate Change 2014: Impacts, Adaptation, and Vulnerability. Part A: Global and Sectoral Aspects. Contribution of Working Group II to the Fifth Assessment Report of the Intergovernmental Panel on Climate Change*; Field, C.B., Barros, V.R., Dokken, D.J., Mach, K.J., Mastrandrea, M.D., Bilir, T.E., Chatterjee, M., Ebi, K.L., Estrada, Y.O., Genova, R.C., et al., Eds.; Cambridge University Press: Cambridge UK; New York, NY, USA, 2014; pp. 485–533.

43. Antle, J.M.; Capalbo, S.M. Adaptation of agricultural and food systems to climate change: An economic and policy perspective. *Appl. Econ. Perspect. Policy* **2010**, *32*, 386–416. [CrossRef]

44. Schmidhuber, J.; Tubiello, F.N. Global food security under climate change. *Proc. Natl. Acad. Sci. USA* **2007**, *104*, 19703–19708. [CrossRef] [PubMed]

45. Attavanich, W.; McCarl, B.A.; Ahmedov, Z.; Fuller, S.W.; Vedenov, D.V. Effects of climate change on US grain transport. *Nat. Clim. Chang.* **2013**, *3*, 638. [CrossRef]

46. Butt, T.A.; McCarl, B.A.; Angerer, J.; Dyke, P.T.; Stuth, J.W. The economic and food security implications of climate change in Mali. *Clim. Chang.* **2005**, *68*, 355–378. [CrossRef]

47. Ahmed, S.A.; Diffenbaugh, N.S.; Hertel, T.W. Climate volatility deepens poverty vulnerability in developing countries. *Environ. Res. Lett.* **2009**, *4*, 034004. [CrossRef]

48. McCarl, B.A.; Hertel, T.W. Climate change as an agricultural economics research topic. *Appl. Econ. Perspect. Policy* **2018**, *40*, 60–78. [CrossRef]

49. Kjellstrom, T.; Kovats, R.S.; Lloyd, S.J.; Holt, T.; Tol, R.S.J. The direct impact of climate change on regional labor productivity. *Arc. Environ. Occup. Health* **2009**, *64*, 217–227. [CrossRef]

50. Hertel, T.W.; Rosch, S.D. *Climate Change, Agriculture and Poverty*; Policy Working Paper; The World Bank: Washington, DC, USA, 2010.

51. Parmesan, C.; Yohe, G. A globally coherent fingerprint of climate change impacts across natural systems. *Nature* **2003**, *421*, 37–42. [CrossRef]

52. Lawing, A.M.; Polly, P.D.; Hews, D.K.; Martins, E.P. Including fossils in phylogenetic climate reconstructions: A deep time perspective on the climatic niche evolution and diversification of spiny lizards (*Sceloporus*). *Am. Nat.* **2016**, *188*, 133–148. [CrossRef]

53. Bellard, C.; Bertelsmeier, C.; Leadley, P.; Thuiller, W.; Courchamp, F. Impacts of climate change on the future of biodiversity: Biodiversity and climate change. *Ecol. Lett.* **2012**, *15*, 365–377. [CrossRef]

54. Urban, M.C.; Bocedi, G.; Hendry, A.P.; Mihoub, J.B.; Peer, G.; Singer, A.; Bridle, J.R.; Crozier, L.G.; De Meester, L.; Godsoe, W.; et al. Improving the forecast for biodiversity under climate change. *Science* **2016**, *353*, aad8466. [CrossRef] [PubMed]

55. Anderson, R.P. Harnessing the world's biodiversity data: Promise and peril in ecological niche modeling of species distributions. *Ann. N. Y. Acad. Sci.* **2012**, *1260*, 66–80. [CrossRef] [PubMed]

56. Amatulli, G.; Domisch, S.; Tuanmu, M.N.; Parmentier, B.; Ranipeta, A.; Malczyk, J.; Jetz, W. A suite of global, cross-scale topographic variables for environmental and biodiversity modeling. *Sci. Data* **2018**, *5*, 180040. [CrossRef] [PubMed]

57. Fick, S.E.; Hijmans, R.J. WorldClim 2: New 1-km spatial resolution climate surfaces for global land areas. *Int. J. Climatol.* **2017**, *37*, 4302–4315. [CrossRef]

58. Hengl, T.; de Jesus, J.M.; Heuvelink, G.B.; Gonzalez, M.R.; Kilibarda, M.; Blagotić, A.; Shangguan, W.; Wright, M.N.; Geng, X.; Bauer-Marschallinger, B.; et al. SoilGrids250 m: Global gridded soil information based on machine learning. *PLoS ONE* **2017**, *12*, e0169748. [CrossRef]

59. Harris, R.M.; Grose, M.R.; Lee, G.; Bindoff, N.L.; Porfirio, L.L.; Fox-Hughes, P. Climate projections for ecologists. *Wiley Interdiscip. Rev. Clim. Chang.* **2014**, *5*, 621–637. [CrossRef]

60. Bagchi, R.; Hole, D.G.; Butchart, S.H.M.; Collingham, Y.C.; Fishpool, L.D.; Plumptre, A.J.; Owiunji, I.; Mugabe, H.; Willis, S.G. Forecasting potential routes for movement of endemic birds among important sites for biodiversity in the Albertine Rift under projected climate change. *Ecography* **2018**, *41*, 401–413. [CrossRef]

61. Thomas, C.D.; Franco, A.M.; Hill, J.K. Range retractions and extinction in the face of climate warming. *Trends Ecol. Evol.* **2006**, *21*, 415–416. [CrossRef]

62. Telwala, Y.; Brook, B.W.; Manish, K.; Pandit, M.K. Climate-induced elevational range shifts and increase in plant species richness in a Himalayan biodiversity epicentre. *PLoS ONE* **2013**, *8*, e57103. [CrossRef]

63. Bell, D.M.; Bradford, J.B.; Lauenroth, W.K. Early indicators of change: Divergent climate envelopes between tree life stages imply range shifts in the western United States: Early indications of tree range shift. *Glob. Ecol. Biogeogr.* **2014**, *23*, 168–180. [CrossRef]

64. Freeman, B.G.; Freeman, A.M.C. Rapid upslope shifts in New Guinean birds illustrate strong distributional responses of tropical montane species to global warming. *Proc. Natl. Acad. Sci. USA* **2014**, *111*, 4490–4494. [CrossRef] [PubMed]

65. Röpke, C.P.; Amadio, S.; Zuanon, J.; Ferreira, E.J.G.; de Deus, C.P.; Pires, T.H.S.; Winemiller, K.O. Simultaneous abrupt shifts in hydrology and fish assemblage structure in a floodplain lake in the central Amazon. *Sci. Rep.* **2017**, *7*, 40170. [CrossRef] [PubMed]

66. Moritz, C.; Patton, J.L.; Conroy, C.J.; Parra, J.L.; White, G.C.; Beissinger, S.R. Impact of a century of climate change on small-mammal communities in Yosemite National Park, USA. *Science* **2008**, *322*, 261–264. [CrossRef] [PubMed]

67. Sgrò, C.M.; Lowe, A.J.; Hoffmann, A.A. Building evolutionary resilience for conserving biodiversity under climate change: Conserving biodiversity under climate change. *Evol. Appl.* **2011**, *4*, 326–337. [CrossRef]

68. Foster, C.R.; Amos, A.F.; Fuiman, L.A. Phenology of six migratory coastal birds in relation to climate change. *Wilson J. Ornithol.* **2010**, *122*, 116–125. [CrossRef]

69. Cameron, G.N.; Scheel, D. Getting warmer: Effect of global climate change on distribution of rodents in Texas. *J. Mammal.* **2001**, *82*, 652–680. [CrossRef]

70. Dantas-Torres, F.; Chomel, B.B.; Otranto, D. Ticks and tick-borne diseases: A One Health perspective. *Trends Parasitol.* **2012**, *28*, 437–446. [CrossRef]

71. Feria-Arroyo, T.P.; Castro-Arellano, I.; Gordillo-Perez, G.; Cavazos, A.L.; Vargas-Sandoval, M.; Grover, A.; Torres, J.; Medina, R.F.; Pérez de León, A.A.; Esteve-Gassent, M.D. Implications of climate change on the distribution of the tick vector *Ixodes scapularis* and risk for Lyme disease in the Texas-Mexico transboundary region. *Parasit. Vectors* **2014**, *7*, 199. [CrossRef]

72. Pérez de León, A.A.; Teel, P.D.; Auclair, A.N.; Messenger, M.T.; Guerrero, F.D.; Schuster, G.; Miller, R.J. Integrated strategy for sustainable cattle fever tick eradication in USA is required to mitigate the impact of global change. *Front. Physiol.* **2012**, *3*, 195. [CrossRef]

73. Castellanos, A.A.; Medeiros, M.C.I.; Hamer, G.L.; Teel, P.D.; Hamer, S.A.; Eubanks, M.D.; Morrow, M.E.; Light, J.E. Decreased small mammal and on-host tick abundance in association with invasive red imported fire ants (*Solenopsis invicta*). *Biol. Lett.* **2016**, *12*, 20160463. [CrossRef]

74. Grime, J.P. Evidence for the existence of three primary strategies in plants and its relevance to ecological and evolutionary theory. *Am. Nat.* **1977**, *111*, 1169–1194. [CrossRef]

75. Huston, M.A. *Biological Diversity: The Coexistence of Species*; Cambridge University Press: Cambridge, UK, 1994.

76. Allen, C.D.; Macalady, A.K.; Chenchouni, H.; Bachelet, D.; McDowell, N.; Vennetier, M.; Kitzberger, T.; Rigling, A.; Breshears, D.D.; Hogg, E.H.; et al. A global overview of drought and heat-induced tree mortality reveals emerging climate change risks for forests. *For. Ecol. Manag.* **2010**, *259*, 660–684. [CrossRef]

77. Aragón-Gastélum, J.L.; Flores, J.; Yáñez-Espinosa, L.; Badano, E.; Ramírez-Tobías, H.M.; Rodas-Ortíz, J.P.; González-Salvatierra, C. Induced climate change impairs photosynthetic performance in *Echinocactus platyacanthus*, an especially protected Mexican cactus species. *Flora-Morphol. Distrib. Funct. Ecol. Plants* **2014**, *209*, 499–503. [CrossRef]

78. Islam, A.; Ahuja, L.R.; Garcia, L.A.; Ma, L.; Saseendran, A.S. Modeling the effect of elevated CO_2 and climate change on reference evapotranspiration in the semi-arid Central Great Plains. *Trans. ASABE* **2012**, *55*, 2135–2146. [CrossRef]

79. Hernandez, E.A.; Uddameri, V. Standardized precipitation evaporation index (SPEI)-based drought assessment in semi-arid South Texas. *Environ. Earth Sci.* **2014**, *71*, 2491–2501. [CrossRef]

80. Heisler-White, J.L.; Blair, J.M.; Kelly, E.F.; Harmoney, K.; Knapp, A.K. Contingent productivity responses to more extreme rainfall regimes across a grassland biome. *Glob. Chang. Biol.* **2009**, *15*, 2894–2904. [CrossRef]

81. Serra-Diaz, J.M.; Maxwell, C.; Lucash, M.S.; Scheller, R.M.; Laflower, D.M.; Miller, A.D.; Tepley, A.J.; Epstein, H.E.; Anderson-Teixeira, K.J.; Thompson, J.R. Disequilibrium of fire-prone forests sets the stage for a rapid decline in conifer dominance during the 21st century. *Sci. Rep.* **2018**, *8*, 6749. [CrossRef]

82. Young, A.M.; Higuera, P.E.; Duffy, P.A.; Hu, F.S. Climatic thresholds shape northern high-latitude fire regimes and imply vulnerability to future climate change. *Ecography* **2017**, *40*, 606–617. [CrossRef]

83. Allen, C.D.; Breshears, D.D. Drought-induced shift of a forest-woodland ecotone: Rapid landscape response to climate variation. *Proc. Natl. Acad. Sci. USA* **1998**, *95*, 14839–14842. [CrossRef]

84. Breshears, D.D.; Cobb, N.S.; Rich, P.M.; Price, K.P.; Allen, C.D.; Balice, R.G.; Romme, W.H.; Kastens, J.H.; Floyd, M.L.; Belnap, J.; et al. Regional vegetation die-off in response to global-change-type drought. *Proc. Natl. Acad. Sci. USA* **2005**, *102*, 15144–15148. [CrossRef]

85. Richter, B.D. Ecologically sustainable water management: Managing river flows for ecological integrity. *Ecol. Appl.* **2003**, *13*, 206–224. [CrossRef]

86. Perkin, J.S.; Gido, K.B.; Costigan, K.H.; Daniels, M.D.; Johnson, E.R. Fragmentation and drying ratchet down Great Plains stream fish diversity. *Aquat. Conserv. Mar. Freshw. Ecosyst.* **2015**, *25*, 639–655. [CrossRef]

87. Postel, S.; Carpenter, S. Freshwater ecosystem services. In *Nature's Services: Societal Dependence on Natural Ecosystems*; Daily, G.C., Ed.; Island Press: Washington, DC, USA, 1997; pp. 195–214. ISBN 9781559634762.

88. Durham, B.W.; Wilde, G.R. Influence of stream discharge on reproductive success of a prairie stream fish assemblage. *Trans. Am. Fish Soc.* **2006**, *135*, 1644–1653. [CrossRef]

89. Rypel, A.L.; Haag, W.R.; Findlay, R.H. Pervasive hydrologic effects on freshwater mussels and riparian trees in southeastern floodplain ecosystems. *Wetlands* **2009**, *29*, 497–504. [CrossRef]

90. Roach, K.A. Texas water wars: How politics and scientific uncertainty influence environmental flow decision-making in the Lone Star state. *Biodivers. Conserv.* **2013**, *22*, 545–565. [CrossRef]

91. Roelke, D.L.; Brooks, B.W.; Grover, J.P.; Gable, G.M.; Schwierzke-Wade, L.; Hewitt, N.C. Anticipated human population and climate change effects on algal blooms of a toxic haptophyte in the south-central USA. *Can. J. Fish. Aquat. Sci.* **2012**, *69*, 1389–1404. [CrossRef]

92. Roelke, D.L.; Barkoh, A.; Brooks, B.W.; Grover, J.P.; Hambright, K.D.; La Claire, J.W., II; Moeller, P.D.R.; Patino, R. A chronicle of a killer alga in the west: Ecology, assessment and management of *Prymnesium parvum* blooms. *Hydrobiologia* **2016**, *764*, 29–50. [CrossRef]

93. Schriever, C.A.; Liess, M. Mapping ecological risk of agricultural pesticide runoff. *Sci. Total Environ.* **2007**, *384*, 264–279. [CrossRef]

94. Fulton, M.H.; Moore, D.W.; Wirth, E.F.; Chandler, G.T.; Key, P.B.; Daugomah, J.W.; Strozier, E.D.; Devane, J.; Clark, J.R.; Lewis, M.A.; et al. Assessment of risk reduction strategies for the management of agricultural nonpoint source pesticide runoff in estuarine ecosystems. *Toxicol. Ind. Health* **1999**, *15*, 201–214. [CrossRef]

95. Derner, J.D.; Johnson, H.B.; Kimball, B.A.; Pinter, P.J.; Polley, H.W.; Tischler, C.R.; Boutton, T.W.; LaMorte, R.L.; Wall, G.W.; Adam, N.R.; et al. Above-and below-ground responses of C3–C4 species mixtures to elevated CO_2 and soil water availability. *Glob. Chang. Biol.* **2003**, *9*, 452–460. [CrossRef]

96. Power, A.G. Ecosystem services and agriculture: Tradeoffs and synergies. *Philos. Trans. R. Soc. B Biol. Sci.* **2010**, *365*, 2959–2971. [CrossRef] [PubMed]

97. Nelson, E.; Mendoza, G.; Regetz, J.; Polasky, S.; Tallis, H.; Cameron, D.R.; Chan, K.M.; Daily, G.C.; Goldstein, J.; Kareiva, P.M.; et al. Modeling multiple ecosystem services, biodiversity conservation, commodity production, and tradeoffs at landscape scales. *Front. Ecol. Environ.* **2009**, *7*, 4–11. [CrossRef]

98. Sandhu, H.S.; Wratten, S.D.; Cullen, R. From poachers to gamekeepers: Perceptions of farmers towards ecosystem services on arable farmland. *Int. J. Agric. Sustain.* **2007**, *5*, 39–50. [CrossRef]

99. Costanza, R.; d'Arge, R.; De Groot, R.; Farber, S.; Grasso, M.; Hannon, B.; Limburg, K.; Naeem, S.; O'Neill, R.V.; Paruelo, J.; et al. The value of the world's ecosystem services and natural capital. *Nature* **1997**, *387*, 253–260. [CrossRef]

100. Benton, T.G. Managing farming's footprint on biodiversity. *Science* **2007**, *315*, 341–342. [CrossRef]

101. Tsiafouli, M.A.; Thébault, E.; Sgardelis, S.P.; De Ruiter, P.C.; Van Der Putten, W.H.; Birkhofer, K.; Hemerik, L.; De Vries, F.T.; Bardgett, R.D.; Brady, M.V.; et al. Intensive agriculture reduces soil biodiversity across Europe. *Glob. Chang. Biol.* **2015**, *21*, 973–985. [CrossRef]

102. Dale, V.H.; Polasky, S. Measures of the effects of agricultural practices on ecosystem services. *Ecol. Econ.* **2007**, *64*, 286–296. [CrossRef]

103. Sandhu, H.S.; Wratten, S.D.; Cullen, R. Organic agriculture and ecosystem services. *Environ. Sci. Policy* **2010**, *13*, 1–7. [CrossRef]

104. U.S. Environmental Protection Agency. Climate Change Adaptation Plan. Available online: https://www.epa.gov/sites/production/files/2015-08/documents/adaptationplans2014_508.pdf (accessed on 25 September 2019).

105. Zhang, Y.W.; Hagerman, A.D.; McCarl, B.A. Influence of Climate Factors on Spatial Distribution of Texas Cattle Breeds. *Clim. Chang.* **2013**, *118*, 183–195. [CrossRef]

106. Torell, L.A.; Murugan, S.; Ramirez, O.A. Economics of flexible versus conservative stocking strategies to manage climate variability risk. *Rangel. Ecol. Manag.* **2010**, *63*, 415–425. [CrossRef]

107. Colby, B.G. Economic impacts of water law—State law and water market development in the southwest. *Nat. Resour. J.* **1988**, *28*, 721–749.

108. Falco, S.D.; Adinolfi, F.; Bozzola, M.; Capitanio, F. Crop insurance as a strategy for adapting to climate change. *J. Agric. Econ.* **2014**, *65*, 485–504. [CrossRef]

109. Mendelsohn, R. Efficient adaptation to climate change. *Clim. Chang.* **2000**, *45*, 583–600. [CrossRef]

110. Hanley, N.; Shogren, J.F.; White, B. *Environmental Economics: In Theory an Practice*, 2nd ed.; Palgrave Macmillan: New York, NY, USA, 2007; ISBN 978-033-397-137-6.

111. Magnan, A.K.; Schipper, E.L.; Burkett, M.; Bharwani, S.; Burton, I.; Eriksen, S.; Gemenne, F.; Schaar, J.; Ziervogel, G. Addressing the risk of maladaptation to climate change. Wiley Interdiscip. *Rev. Clim. Chang.* **2016**, *7*, 646–665. [CrossRef]

112. Butt, T.A.; Mccarl, B.A.; Kergna, A.O. Policies for Reducing Agricultural Sector Vulnerability to Climate Change in Mali. *Clim. Policy* **2006**, *5*, 583–598. [CrossRef]

113. Seo, S.N.; McCarl, B.A.; Mendelsohn, R.O. From beef cattle to sheep under global warming? An analysis of adaptation by livestock species choice in South America. *Ecol. Econ.* **2010**, *69*, 2486–2494. [CrossRef]

114. Zhang, Y.W.; McCarl, B.A.; Jones, J.P.H. An Overview of Mitigation and Adaptation Needs and Strategies for the Livestock Sector. *Climate* **2017**, *5*, 95. [CrossRef]

115. Seo, S.N.; Mendelsohn, R.O.; Dinar, A.; Kurukulasuriya, P. Adapting to climate change mosaically: An analysis of African livestock management by agro-ecological zones. *BE J. Econ. Anal. Policy* **2009**, *9*. [CrossRef]

116. Fuhlendorf, S.D.; Engle, D.M. Restoring heterogeneity on rangelands: Ecosystem management based on evolutionary grazing patterns: We propose a paradigm that enhances heterogeneity instead of homogeneity to promote biological diversity and wildlife habitat on rangelands grazed by livestock. *BioScience* **2001**, *51*, 625–632. [CrossRef]

117. Megersa, B.; Markemann, A.; Angassa, A.; Ogutu, J.O.; Piepho, H.P.; Zárate, A.V. Livestock diversification: An adaptive strategy to climate and rangeland ecosystem changes in southern Ethiopia. *Hum. Ecol.* **2014**, *42*, 509–520. [CrossRef]

118. Pequeño-Ledezma, M.; Alanís-Rodríguez, E.; Molina-Guerra, V.M.; Mora-Olivo, A.; Alcalá-Rojas, A.G.; Martínez-Ávalos, J.G.; Garza-Ocañas, F. Plant composition and structure of two post-livestock areas of Tamaulipan thornscrub, Mexico. *Rev. Chil. Hist. Natl.* **2018**, *91*, 4. [CrossRef]

119. Joyce, L.A.; Briske, D.D.; Brown, J.R.; Polley, H.W.; McCarl, B.A.; Bailey, D.W. Climate change and North American rangelands: Assessment of mitigation and adaptation strategies. *Rangel. Ecol. Manag.* **2013**, *66*, 512–528. [CrossRef]

120. Derner, J.D.; Boutton, T.W.; Briske, D.D. Grazing and ecosystem carbon storage in the North American Great Plains. *Plant Soil* **2006**, *280*, 77–90. [CrossRef]

121. Derner, J.D.; Schuman, G.E. Carbon sequestration and rangelands: A synthesis of land management and precipitation effects. *J. Soil Water Conserv.* **2007**, *62*, 77–85.

122. Rocha, J.F.; Martínez, R.; López-Villalobos, N.; Morris, S.T. Tick burden in Bos taurus cattle and its relationship with heat stress in three agroecological zones in the tropics of Colombia. *Parasites Vectors* **2019**, *12*, 73. [CrossRef]

123. Nyamushamba, G.B.; Mapiye, C.; Tada, O.; Halimani, T.E.; Muchenje, V. Conservation of indigenous cattle genetic resources in Southern Africa's smallholder areas: Turning threats into opportunities—A review. *Asian Australas. J. Anim. Sci.* **2017**, *30*, 603. [CrossRef]

124. Adams, R.; McCarl, B.; Segerson, K.; Rosenzweig, C.; Bryant, K.; Dixon, B.; Conner, R.; Evenson, R.; Ojima, D. The economic effect of climate change on US agriculture. In *The Economic Impact of Climate Change on the United States Economy*; Mendelsohn, R., Neumann, J., Eds.; Cambridge University Press: Cambridge, UK, 1999.

125. Chen, C.C.; McCarl, B.A.; Chang, C.C. Climate Change, Sea Level Rise and Rice: Global Market Implications. *Clim. Chang.* **2012**, *110*, 543–560. [CrossRef]

126. Karlen, D.L.; Cambardella, C.A.; Kovar, J.L.; Colvin, T.S. Soil quality response to long-term tillage and crop rotation practices. *Soil Till. Res.* **2013**, *133*, 54–64. [CrossRef]

127. Page, K.; Dang, Y.; Dalal, R. Impacts of conservation tillage on soil quality, including soil-borne crop diseases, with a focus on semi-arid grain cropping systems. *Australas. Plant Pathol.* **2013**, *42*, 363–377. [CrossRef]

128. Barbet-Massin, M.; Thuiller, W.; Jiguet, F. The fate of European breeding birds under climate, land-use and dispersal scenarios. *Glob. Chang. Biol.* **2012**, *18*, 881–890. [CrossRef]

129. Mantyka-Pringle, C.S.; Visconti, P.; Di Marco, M.; Martin, T.G.; Rondinini, C.; Rhodes, J.R. Climate change modifies risk of global biodiversity loss due to land-cover change. *Biol. Conserv.* **2015**, *187*, 103–111. [CrossRef]

130. Brosi, B.J.; Briggs, H.M. Single pollinator species losses reduce floral fidelity and plant reproductive function. *Proc. Natl. Acad. Sci. USA* **2013**, *110*, 13044–13048. [CrossRef] [PubMed]

131. Allen-Wardell, G.; Bernhardt, P.; Bitner, R.; Burquez, A.; Buchmann, S.; Cane, J.; Cox, P.A.; Dalton, V.; Feinsinger, P.; Ingram, M.; et al. The Potential Consequences of Pollinator Declines on the Conservation of Biodiversity and Stability of Food Crop Yields. *Conserv. Biol.* **1998**, *12*, 8–17.

132. Rathcke, B.J.; Jules, E.S. Habitat fragmentation and plant-pollinator interactions. *Curr. Sci.* **1993**, *65*, 273–277.

133. Martin, J.M.; Mead, J.I.; Barboza, P.S. Bison body size and climate change. *Ecol. Evol.* **2018**, *8*, 4564–4574. [CrossRef] [PubMed]

134. Torell, G.; Lee, K. Impact of Climate Change on Livestock Returns and Rangeland Ecosystem Sustainability in the Southwest. *Agric. Resour. Econ. Rev.* **2018**, *47*, 336–356. [CrossRef]

135. Geruo, A.; Velicogna, I.; Kimball, J.S.; Du, J.; Kim, Y.; Colliander, A.; Njoku, E. Satellite-observed changes in vegetation sensitivities to surface soil moisture and total water storage variations since the 2011 Texas drought. *Environ. Res. Lett.* **2017**, *12*, 054006. [CrossRef]

136. Schwantes, A.M.; Swenson, J.J.; González-Roglich, M.; Johnson, D.M.; Domec, J.C.; Jackson, R.B. Measuring canopy loss and climatic thresholds from an extreme drought along a fivefold precipitation gradient across Texas. *Glob. Chang. Biol.* **2017**, 5120–5135. [CrossRef]

137. McDonald, R.I.; Girvetz, E.H. Two Challenges for USA Irrigation Due to Climate Change: Increasing Irrigated Area in Wet States and Increasing Irrigation Rates in Dry States. *PLoS ONE* **2001**, *8*, e65589. [CrossRef]

138. Rodríguez-Díaz, J.A.; Weatherhead, E.K.; Knox, J.W.; Camacho, E. Climate change impacts on irrigation water requirements in the Guadalquivir river basin in Spain. *Reg. Environ. Chang.* **2007**, *7*, 149–159. [CrossRef]

139. Mainali, K.P.; Warren, D.L.; Dhileepan, K.; McConnachie, A.; Strathie, L.; Hassan, G.; Karki, D.; Shrestha, B.B.; Parmesan, C. Projecting future expansion of invasive species: Comparing and improving methodologies for species distribution modeling. *Glob. Chang. Biol.* **2015**, *21*, 4464–4480. [CrossRef] [PubMed]

140. Burlakova, L.E.; Karatayev, A.Y.; Karatayev, V.A.; May, M.E.; Bennett, D.L.; Cook, M.J. Biogeography and conservation of freshwater mussels (Bivalvia: Unionidae) in Texas: Patterns of diversity and threats. *Divers. Distrib.* **2011**, *17*, 393–407. [CrossRef]

141. Wolfe, D.W.; Ziska, L.; Petzoldt, C.; Seaman, A.; Chase, L.; Hayhoe, K. Projected change in climate thresholds in the Northeastern, USA: Implications for crops, pests, livestock, and farmers. *Mitig. Adapt. Strateg. Glob. Chang.* **2008**, *13*, 555–575. [CrossRef]

142. Smith, R.G.; Menalled, F.D. *Integrated Strategies for Managing Agricultural weeds: Making Cropping Systems Less Susceptible to Weed Colonization and Establishment Department of Land Resources and Environmental Sciences*; Montana State University Extension: Bozeman, MT, USA, 2006.

143. Lehikoinen, A.; Jaatinen, K. Delayed autumn migration in northern European waterfowl. *J. Ornithol.* **2012**, *153*, 563–570. [CrossRef]

144. Schummer, M.L.; Coluccy, J.M.; Mitchell, M.; Van Den Elsen, L. Long-term trends in weather severity indices for dabbling ducks in eastern North America. *Wildl. Soc. Bull.* **2017**, *41*, 615–623. [CrossRef]

145. Rand, P.S.; Hinch, S.G.; Morrison, J.; Foreman, M.G.G.; MacNutt, M.J.; MacDonald, J.S.; Healey, M.C.; Farrell, A.P.; Higgs, D.A. Effects of river discharge, temperature, and future climates on energetics and mortality of adult migrating Fraser River sockeye salmon. *Trans. Am. Fish. Soc.* **2016**, *135*, 655–667. [CrossRef]

146. MacFarlane, R.B. Energy dynamics and growth of Chinook salmon (*Oncorhynchus tshawytscha*) from the Central Valley of California during the estuarine phase and first ocean year. *Can. J. Fish. Aquat. Sci.* **2010**, *67*, 1549–1565. [CrossRef]

147. Todd, C.D.; Hughes, S.L.; Marshall, C.T.; MacLean, J.C.; Lonergan, M.E.; Biuw, E.M. Detrimental effects of recent ocean surface warming on growth condition of Atlantic salmon. *Glob. Chang. Biol.* **2008**, *14*, 958–970. [CrossRef]

148. Sheridan, J.A.; Bickford, D. Shrinking body size as an ecological response to climate change. *Nat. Clim. Chang.* **2011**, *1*, 401. [CrossRef]

149. Jonsson, B.; Jonsson, N. Factors affecting marine production of Atlantic salmon (Salmo salar). *Can. J. Fish. Aquat. Sci.* **2004**, *61*, 2369–2383. [CrossRef]

150. Tscharntke, T.; Batáry, P.; Dormann, C.F. Set-aside management: How do succession, sowing patterns and landscape context affect biodiversity? *Agric. Ecosyst. Environ.* **2011**, *143*, 37–44. [CrossRef]

151. Palm, C.; Blanco-Canqui, H.; DeClerck, F.; Gatere, L.; Grace, P. Conservation agriculture and ecosystem services: An overview. *Agric. Ecosyst. Environ.* **2014**, *187*, 87–105. [CrossRef]

152. Busari, M.A.; Kukal, S.S.; Kaur, A.; Bhatt, R.; Dulazi, A.A. Conservation tillage impacts on soil, crop and the environmentInt. *Soil Water Conserv. Res.* **2015**, *3*, 119–129. [CrossRef]

153. Seufert, V.; Ramankutty, N.; Foley, J.A. Comparing the yields of organic and conventional agriculture. *Nature* **2012**, *485*, 229. [CrossRef] [PubMed]

154. Gebbers, R.; Adamchuk, V.I. Precision agriculture and food security. *Science* **2010**, *327*, 828–831. [CrossRef]

155. Bongiovanni, R.; Lowenberg-DeBoer, J. Precision agriculture and sustainability. *Precis. Agric.* **2004**, *5*, 359–387. [CrossRef]

156. Lipper, L.; Thornton, P.; Campbell, B.M.; Baedeker, T.; Braimoh, A.; Bwalya, M.; Caron, P.; Cattaneo, A.; Garrity, D.; Henry, K.; et al. Climate-smart agriculture for food security. *Nat. Clim. Chang.* **2014**, *4*, 1068–1072. [CrossRef]

157. Campbell, B.; Thornton, P.; Zougmore, D.; van Asten, P.; Lipper, L. Sustainable intensification: What is its role in climate smart agriculture? *Curr. Opin. Environ. Sustain.* **2014**, *8*, 39–43. [CrossRef]

158. Thorpe, A.S.; Barnett, D.T.; Elmendorf, S.C.; Hoekman, D.; Jones, K.D.; Levan, K.E.; Stanish, L.F. Introduction to the sampling designs of the National Ecological Observatory Network Terrestrial Observation System. *Ecosphere* **2016**, *7*, e01627. [CrossRef]

159. Trivedi, P.; Delgado-Baquerizo, M.; Anderson, I.C.; Singh, B.K. Response of soil properties and microbial communities to agriculture: Implications for primary productivity and soil health indicators. *Front. Plant Sci.* **2016**, *7*, 990. [CrossRef]

160. Baltensperger, A.P.; Huettmann, F. Predicted shifts in small mammal distributions and biodiversity in the altered future environment of Alaska: An open access data and machine learning perspective. *PLoS ONE* **2018**, *13*, e0194377. [CrossRef] [PubMed]

161. Beck, J.; Sieber, A. Is the spatial distribution of mankind's most basic economic traits determined by climate and soil alone? *PLoS ONE* **2010**, *5*, e10416. [CrossRef] [PubMed]

162. Davis, A.P.; Gole, T.W.; Baena, S.; Moat, J. The impact of climate change on indigenous Arabica coffee (*Coffea arabica*): Predicting future trends and identifying priorities. *PLoS ONE* **2012**, *7*, e47981. [CrossRef] [PubMed]

163. Curtis-Robles, R.; Wozniak, E.J.; Auckland, L.D.; Hamer, G.L.; Hamer, S.A. Combining public health education and disease ecology research: Using citizen science to assess Chagas disease entomological risk in Texas. *PLoS Negl. Trop. Dis.* **2015**, *9*, e0004235. [CrossRef]

164. Meierhofer, M.B.; Johnson, J.S.; Leivers, S.J.; Pierce, B.L.; Evans, J.E.; Morrison, M.L. Winter habitats of bats in Texas. *PLoS ONE* **2019**, *14*, e0220839. [CrossRef]

165. Ikard, W. Encouraging Conservation in the Lone Star State: How Texas Can Improve Incentives for Landowners to Preserve Private Property from Development. *Tex. Envtl. LJ* **2008**, *39*, 147–166.

166. Sorice, M.G.; Haider, W.; Conner, J.R.; Ditton, R.B. Incentive structure of and private landowner participation in an endangered species conservation program. *Conserv. Biol.* **2011**, *25*, 587–596. [CrossRef]

167. Kreuter, U.P.; Wolfe, D.W.; Hays, K.B.; Conner, J.R. Conservation credits—Evolution of a market-oriented approach to recovery of species of concern on private land. *Rangel. Ecol. Manag.* **2017**, *70*, 264–272. [CrossRef]

168. Agrawal, A. *The Role of Local Institutions in Adaptation to Climate Change*; World Bank: Washington, DC, USA, 2008. [CrossRef]

169. Marshall, N.A.; Smajgl, A. Understanding variability in adaptive capacity on rangelands. *Rangel. Ecol. Manag.* **2013**, *66*, 88–94. [CrossRef]

170. Polasky, S.; Carpenter, S.R.; Folke, C.; Keeler, B. Decision-making under great uncertainty: Environmental management in an era of global change. *Trends Ecol. Evol.* **2011**, *26*, 398–404. [CrossRef]

MDPI
St. Alban-Anlage 66
4052 Basel
Switzerland
Tel. +41 61 683 77 34
Fax +41 61 302 89 18
www.mdpi.com

Climate Editorial Office
E-mail: climate@mdpi.com
www.mdpi.com/journal/climate

www.ingramcontent.com/pod-product-compliance
Lightning Source LLC
LaVergne TN
LVHW070652170726
843515LV00004B/393